Praise for *Why Does E=mc²?*

"To get at the origins of E=mc², the poster-child for Einsteins's special theory of relativity, [Cox and Forshaw] must delve into deep principles of science and wield a good deal of mathematics. They do it well . . . They have blazed a clear trail into forbidding territory, from the mathematical structure of space-time all the way to atom bombs, astrophysics and the origin of mass."

—New Scientist

"Many authors have tried to explain [the equation's] origins, with mixed results. It's hard to think of two authors more qualified for the job than Brian Cox and Jeff Forshaw. . . . They do a grand job of answering the question in the book's title, and of tying it to the cutting edge of 21st century physics."

—BBC Focus Magazine

"[*Why Does E=mc²?*] is clear, sparkling in places, and totally without vanity . . . anyone with an adventurous mind should be intrigued by what two smart physicists say about it in plain language . . . [A] delightful little book."

—The Huffington Post

"To move beyond a cursory understanding of Einstein's iconic equation, put yourself in the adept hands of physicists and science educators Brian Cox and Jeff Forshaw. Using clear language and a few clearly explained equations, they demystify physics' most counterintuitive claims."

—*Seed*

"A mild-mannered, digressive, mostly math-free walk-through of the world's most famous equation . . . [that] reminds us that Einstein's equation is not some esoteric idea best pondered by scientific supermen, but a profound insight that continues to change lives . . . Cox and Forshaw's enthusiasm for their material is plain . . . You will find them accommodating escorts."

—*Boston Globe*

"There is a great deal of knowledge and quite competent explanation throughout [*Why Does E=mc²?*], which should serve as a dream come true for anyone who ever loved science, or wanted to learn more about it without having to go back to school. Come to this read with an open mind and a desire to learn, and you will come away with a treasure trove of knowledge."

—*Sacramento Book Review*

"Cox and Forshaw offer lay readers a fascinating account of modern scientists' view of the world, and how it got that way. . . . [They] provide the historical context that set the stage for Einstein's discovery . . . [and] also clearly explain the tide shift that Einstein caused, transforming scientists' understanding of the world. . . . Though the basics are covered in detail, there's plenty here for science buffs to ponder."

—*Publishers Weekly*

"Pairs the enthusiasm of newcomers with the knowledge of experts . . . Cox and Forshaw have aimed their tour of gravity, mass and quantum weirdness squarely at the math-shy general public. Readers in this category should benefit from plenty of helpful and mostly non-mathematical explanations."

—*Physics World*

"I can think of no one, Stephen Hawking included, who more perfectly combines authority, knowledge, passion, clarity and powers of elucidation than Brian Cox. If you really want to know how Big Science works and why it matters to each of us in the smallest way then be entertained by this dazzlingly enthusiastic man. Can someone this charming really be a professor?"

—Stephen Fry

"Cox and Forshaw take the equation that all of us know and few of us understand—and make it crystal clear for all of us. A thrilling experience of passionate comprehension."

—Ann Druyan, co-writer, *Cosmos* television series

Why Does E=mc^2

Why Does E=mc²

(And Why Should We Care?)

BRIAN COX AND JEFF FORSHAW

DA CAPO PRESS

A MEMBER OF THE PERSEUS BOOKS GROUP

Set in Minion Pro by the Perseus Books Group

Printed in the UK by CPI Clowes Beccles NR34 7TL

Library of Congress Cataloging-in-Publication Data

Cox, Brian, 1968-
    Why does e=mc2 : (and why should we care?) / Brian Cox and Jeff Forshaw.
        p. cm.
        Includes index.
    ISBN 978-0-306-81758-8 (alk. paper)
1. Einstein field equations. 2. Special relativity (Physics)—Mathematics. 3. Space and time—Mathematics. I. Forshaw, J. R. (Jeffrey Robert), 1968- II. Title.
QC173.6.C68 2009
530.11—dc22

                                            2009009291

First Da Capo Press edition 2009
First Da Capo Press paperback edition 2010
Published by Da Capo Press
A Member of the Perseus Books Group
www.dacapopress.com

Da Capo Press books are available at special discounts for bulk purchases in the United States by corporations, institutions, and other organizations. For more information, please contact the Special Markets Department at the Perseus Books Group, 2300 Chestnut Street, Suite 200, Philadelphia, PA 19103, or call (800) 810-4145, ext. 5000, or e-mail special.markets@perseusbooks.com.

UK ISBN: 978-0-306-81911-7

10 9 8 7 6 5 4 3 2 1

To our families, especially Gia, Mo, George, David, Barbara, Sandra, Naomi, Isabel, Sylvia, Thomas, and Michael

CONTENTS

ACKNOWLEDGMENTS

We thank our management and agents, Susan, Diane, and George, and our editors, Ben and Cisca. Of our scientific colleagues, we should particularly like to thank Richard Battye, Fred Loebinger, Robin Marshall, Simone Marzani, Ian Morison, and Gavin Smith. Special thanks to Naomi Baker, especially for her comments on the early chapters, and to Gia Milinovich for asking the question.

PREFACE

Our aim in this book is to describe Einstein's theory of space and time in the simplest way we can while at the same time revealing its profound beauty. Ultimately, this will allow us to arrive at his famous equation $E = mc^2$ using mathematics no more complicated than Pythagoras' theorem. And don't worry if you can't remember Pythagoras, because we will describe that as well. Equally important, we want every reader who finishes this little book to see how modern physicists think about nature and build theories that become profoundly useful and ultimately change our lives. By building a model of space and time, Einstein paved the way for an understanding of how stars shine, uncovered the deep reason why electric motors and generators work, and ultimately laid the foundation on which all of modern physics rests. This book is also intended to be provocative and challenging. The physics itself is not at issue: Einstein's theories are very well established and backed up by a great deal of experimental evidence, as we shall discover as the book unfolds. In due course, it is very important to emphasize, Einstein may be forced to give way to an even more accurate picture of nature. In science, there are no universal truths, just views of the world that have yet to be shown to be false. All we can say for certain is that, for now, Einstein's theory works. Instead, the provocation lies in the way the science challenges us to think about the world around us. Scientist or not, each of us has intuition and we all infer things about the world from our everyday experiences. If we subject our observations to the cold and precise light of the scientific method, however, we often discover that nature confounds our intuition. As this book unfolds, we will discover that when things whiz about at high speeds, common-sense notions regarding space and time are dashed and replaced by something entirely new, unexpected, and elegant. The lesson is a salutary and humbling one, and it leaves many scientists with a sense of awe: The universe is much richer than our everyday experiences would have us believe. Perhaps most wonderful of all is the fact that the new physics, for all its richness, is filled with a breathtaking mathematical elegance.

Difficult as it may sometimes seem, science at its heart is not a complicated discipline. One might venture to say that it is an attempt at removing our innate prejudices in order to observe the world as objectively as possible. It may be more or less successful in that goal but few can doubt its success in teaching us how the universe "works." The really difficult thing is to learn not to trust what we might like to think of as common sense. By teaching us to accept nature for what it is, and not for what our prejudice may suggest that it should be, the scientific method has delivered the modern technological world. In short, it works.

In the first half of the book we will derive the equation $E = mc^2$. By "derive," we mean that we will show how Einstein reached the conclusion that energy is equal to mass multiplied by the speed of light squared, which is what the equation says.

Think about this for a moment and it seems like a very odd thing. Perhaps the most familiar kind of energy is the energy of motion; if someone throws a cricket ball at your face, then it hurts when it hits you. A physicist would say that this is because the cricket ball was given energy by the thrower, and this energy is transferred to your face when your face stops the ball. Mass is a measure of how much stuff an object contains. A cricket ball is more massive than a table-tennis ball, but less massive than a planet. What $E = mc^2$ says is that energy and mass are interchangeable much like dollars and euros are interchangeable, and that the speed of light squared is the exchange rate. How on earth could Einstein have reached this conclusion, and how could the speed of light find its way into an equation about the relationship between energy and mass? We do not assume any prior scientific knowledge and we avoid mathematics as much as possible. Nevertheless, we do aim to offer the reader a genuine explanation (and not merely a description) of the science. In that regard especially, we hope to offer something new.

In the latter parts of the book, we will see how $E = mc^2$ underpins our understanding of the workings of the universe. Why do stars shine? Why is nuclear power so much more efficient than coal or oil? What is mass? This question will lead us into the world of modern particle physics, the Large Hadron Collider at CERN in Geneva, and the hunt for the Higgs particle that may lead to an explanation for the very origin of mass. The book finishes with Einstein's remarkable discovery that the structure of space and time is ultimately responsible for the force of gravity and the strange idea that the earth is falling "in a straight line" around the sun.

Space and Time

What do the words "space" and "time" mean to you? Perhaps you picture space as the blackness between the stars as you turn your gaze toward the sky on a cold winter's night. Or maybe you see the void between earth and moon sailed by spacecraft clad in golden foil, bedecked with the stars and stripes, piloted into magnificent desolation by shaven-headed explorers with names like Buzz. Time may be the tick of your watch or the reddening of the leaves as the earth's yearly circuit of the sun tilts northern latitudes toward shade for the 5 billionth time. We all have an intuitive feel for space and time; they are part of the fabric of our existence. We move through space on the surface of our blue world as time ticks by.

During the late years of the nineteenth century, a series of scientific breakthroughs in apparently unrelated fields began to force physicists to reexamine these simple and intuitive pictures of space and time. By the early years of the twentieth century, Albert Einstein's colleague and tutor Hermann Minkowski was moved to write his now-famous obituary for the ancient arena within which planets orbit and great journeys are made: "From henceforth, space by itself, and time by itself, have vanished into the merest shadows and only a kind of blend of the two exists in its own right."

What could Minkowski have meant by a blend of space and time? To understand this almost mystical-sounding statement is to understand Einstein's special theory of relativity—the theory that introduced the world to that most famous of all equations, $E = mc^2$, and placed forever center-stage in our understanding of the fabric of the universe the quantity with the symbol c, the speed of light.

Einstein's special theory of relativity is at its heart a description of space and time. Central to the theory is the notion of a special speed, a speed beyond which nothing in the universe, no matter how powerful, can accelerate. This speed is the speed of light; 299,792,458 meters per second in the vacuum of empty space. Traveling at this speed, a flash of light beamed out from Earth takes eight minutes to pass by the sun, 100,000 years to cross our own Milky Way galaxy, and over 2 million years to reach our nearest galactic neighbor, Andromeda. Tonight, the largest telescopes on Earth will gaze outward into the blackness of space and capture ancient light from distant, long-dead suns at the edge of the observable universe. This light began its journey over 10 billion years ago, several billion years before the earth was formed from a collapsing cloud of interstellar dust. The speed of light is fast, but nowhere near infinitely so. When faced with the great distances between the stars and galaxies, light speed can be frustratingly slow; slow enough that we can accelerate very small objects to within a fraction of a percent of the speed of light with machines like the 27-kilometer Large

Hadron Collider at the European Center for Particle Physics (CERN) in Geneva, Switzerland.

The existence of such a special speed, a cosmic speed limit, is a strange concept. As we will discover later in this book, linking this special speed with the speed of light turns out to be something of a red herring. It has a much deeper role to play in Einstein's universe, and there is a good reason why light travels at the speed it does. We will get to that later on. For now, suffice to say that when objects approach the special speed, strange things happen. How else could an object be prevented from accelerating beyond that speed? It's as though there were a universal law of physics that prevented your car going faster than seventy miles per hour, no matter how large the engine. Unlike a speed restriction, however, this law is not something that needs to be enforced by some kind of ethereal police force. The very fabric of space and time is constructed in such a way that it is absolutely impossible to break the law, and this turns out to be extremely fortunate, for otherwise there would be unpleasant consequences. Later, we shall see that if it were possible to exceed the speed of light, we could construct time machines capable of transporting us backward through history to any point in the past. We could imagine journeying back to a time before we were born and, by accident or design, preventing our parents from ever meeting. This makes for excellent science fiction, but it is no way to build a universe, and indeed Einstein found that the universe is not built like this. Space and time are delicately interwoven in a way that prevents such paradoxes from occurring. However, there is a price to pay: We must jettison our deeply held notions of space and time. Einstein's universe is one in which moving clocks tick slowly, moving objects shrink, and we can journey billions of years into the future. It is a universe in which a human lifetime can be stretched almost indefinitely. We could watch the sun die, the earth's oceans boil away, and our solar system be plunged into perpetual night. We could watch the birth of stars from swirling dust clouds, the formation of planets and maybe the origins of life on new, as yet unformed worlds. Einstein's universe allows us to journey into the far future, while keeping the doors to the past firmly locked behind us.

By the end of this book, we will see how Einstein was forced to such a fantastical picture of our universe, and how this picture has been shown to be correct in many scientific experiments and technological applications. The satellite navigation system in your car, for example, is designed to account for the fact that time ticks at a different rate on the orbiting satellites than it does on the ground. Einstein's picture is radical: Space and time are not what they seem.

But we are getting ahead of ourselves. To understand and appreciate Einstein's radical discovery, we must first think very carefully about the two concepts at the heart of relativity theory, space and time.

Imagine you are reading a book while riding on an aircraft. At 12:00 you glance at your watch, decide to put your book down, leave your seat, and walk down the aisle to chat with your friend ten rows in front of you. At 12:15 you return to your seat, sit down, and pick up your book. Common sense tells you that you have returned to the same place. You had to walk the same ten rows to get back to your seat, and when you returned your book was where you left it. Now think a little more deeply about the concept of "the same place." This might seem a little pedantic, because it's intuitively obvious what we mean when we describe a place. We can call a friend and arrange to meet up for a drink in a bar, and the bar won't have moved by the time we both arrive. It will be in the same place that we left it, quite possibly the night before. Many things in this opening chapter will appear at first sight to be pedantic, but stick with it. Thinking carefully about these apparently obvious concepts will lead us in the footsteps of Aristotle, Galileo Galilei, Isaac Newton, and Einstein. How, then, could we go about defining precisely what we mean by "the same place"? We already know how to do this on the surface of the earth. A globe has a set of grid lines, lines of latitude and longitude, drawn onto its surface. Any place on the earth's surface can be described by two numbers, representing the position on this grid. For example, the city of Manchester in the UK is located at 53 degrees 30 minutes north, and 2 degrees 15 minutes west. These two numbers tell us exactly where to find Manchester, given that we all agree on the locations of the equator and the Greenwich Meridian. Therefore, by simple analogy, one way to pin down the location of any point, whether on the earth's surface or not, would be to picture an imaginary three-dimensional grid, extending upward from the earth's surface and into the air. Indeed, the grid could also carry on downward through the center of the earth and out the other side. We could then describe where everything in the world sits relative to the grid, whether in the air, on the surface, or below ground. In fact, we needn't stop with just the world. The grid could extend outward beyond the moon, past Jupiter, Neptune, and Pluto, beyond even the edge of the Milky Way galaxy to the farthest reaches of the universe. Given our giant, possibly infinitely large, grid we can work out where everything is, which to paraphrase Woody Allen, is very useful if you're the kind of person who can never remember where you put things. Our grid therefore defines an arena within which everything exists, a kind of giant box containing all objects in the universe. We may even be tempted to call this giant arena "space."

Let's get back to the question of what is meant by "the same place" and return to the aircraft example. You might suppose that at 12:00 and 12:15 you were at the same point in space. Now imagine what the sequence of events looks like to a person sitting on the ground watching the plane. If she sees the plane fly overhead at 600 miles per hour, she would say that between 12:00 and 12:15 you had moved 150 miles. In other words, you were at different points in space at 12:00 and 12:15. Who is correct? Who has moved, and who has stood still?

If you can't see the answer to this apparently simple question, then you are in good company. Aristotle, one of the greatest minds of ancient Greece, got it dead wrong. Aristotle would have answered unequivocally that it is you, the passenger on the aircraft, who is moving. Aristotle believed that the earth stands still at the center of the universe. The sun, moon, planets, and stars rotate around the earth attached to fifty-five concentric crystalline spheres, stacked inside each other like Russian dolls. He therefore shared our intuitively satisfying concept of space: the box or arena in which the earth and the spheres are placed. To modern ears, this picture of the universe consisting only of the earth and a set of spinning spheres sounds rather quaint.

But think for a moment about what conclusion you would draw if nobody had told you that the earth rotates around the sun and that the stars are distant suns, some many thousands of times brighter than our nearby star but billions and billions of miles away. It certainly doesn't feel like the earth is adrift in an unimaginably large universe. Our modern worldview was hard-won and is often counterintuitive. If the picture of the universe we have developed through thousands of years of experiment and thought was obvious, then the greats of the past, such as Aristotle, would have worked it out for themselves. This is worth remembering if you find any of the concepts in this book difficult; the greatest minds of antiquity may well have agreed with you.

To find the flaw in Aristotle's answer, let us accept his picture for a moment and see where it leads. According to Aristotle, we should fill space with imaginary grid lines centered on the earth and work out where everything is, and who is doing the moving. If we accept this picture of space as a box filled with objects, with the earth fixed at its center, then it is obvious that you, the passenger on the plane, have changed your position in the box, while the person watching you fly by is standing still on the surface of the earth, hanging motionless in space. In other words, there is such a thing as absolute motion and therefore absolute space. An object is in absolute motion if it changes its position in space, as measured against the imaginary grid fixed to the center of the earth, as time ticks by.

A problem with this picture, of course, is that the earth is not standing motionless at the center of the universe; it is a spinning ball in orbit around the sun. In fact, the earth is moving at about 67,000 miles per hour relative to the sun. If you go to bed at night and sleep for eight hours, you'll have traveled over half a million miles by the time you wake up. You could even claim that, in about 365 days, your bedroom would have returned to exactly the same point in space since the earth would have completed one full orbit around the sun. You might therefore decide to change your picture a little, while keeping the spirit of Aristotle's view intact. Why not center the grid on the sun? It's a simple enough thought, but it's wrong too because the sun itself is in orbit around the center of the Milky Way galaxy. The Milky Way is our local island of over 200,000 million suns, and as you can probably imagine it's very large and takes quite a while to get around. The sun, with the earth in tow, is traveling around the Milky Way at 486,000 miles per hour, at a distance of 156,000 trillion miles from the center. At this speed, it takes 226 million years to complete one orbit. And so, perhaps one more step might be sufficient to save Aristotle. Center the grid at the center of our Milky Way galaxy and you could be led to another evocative thought: As you lie in your bed, imagine what the world looked like the last time the earth was "here" at this very point in space. A dinosaur was grazing in the early morning shadows, eating prehistoric leaves at the place where your bedroom now stands. Wrong again. In fact, the galaxies themselves are racing away from each other, and the more distant the galaxy, the faster it recedes from us. Our motion among the myriad galaxies that make up the universe appears to be extremely difficult indeed to pin down.

So Aristotle has a problem, because it seems to be impossible to define exactly what is meant by the words "standing still." In other words, it seems impossible to work out where to center the imaginary grid against which we can work out where things are, and thereby decide what is standing still and what is moving. Aristotle himself never had to face this problem because his picture of a stationary Earth surrounded by rotating spheres was not seriously challenged for almost 2,000 years. Perhaps it should have been, but as we have already said, these things are far from obvious even to the greatest of minds. Claudius Ptolemaeus, known today as Ptolemy, worked in the great Library of Alexandria in Egypt in the second century. He was a careful observer of the night sky, and he worried about the apparently strange motion through the heavens of the five then-known planets, or "wandering stars," from which the word "planet" is derived. When viewed from Earth over many months, the planets do not follow a smooth path across the starry background, but appear to perform loop-the-loops in the sky. This strange motion is known as retrograde motion and had in fact been known for many thousands of years before Ptolemy. The ancient Egyptians described Mars as the planet "who travels backward." Ptolemy agreed with Aristotle that the planets were rotating around a stationary Earth, but to explain the retrograde motion he was forced to attach them to smaller off-center rotating wheels, which in turn were attached to the spinning spheres. This rather complicated model was able to explain the motion of the planets across the night sky, although it is far from elegant. The true explanation of the retrograde motion of the planets had to wait for the mid-sixteenth century and Nicholas Copernicus, who proposed the more elegant (and more correct) explanation that the earth is not stationary at the center of the universe, but in fact orbits around the sun along with the rest of the planets. Copernicus's work was not without its detractors and was removed from the Catholic Church's banned list only in 1835. Precision measurements by Tycho Brahe, and the work of Johannes Kepler, Galileo, and Newton, finally established not only that Copernicus was correct, but led to a theory of planetary motion in the form of Newton's laws of motion and gravitation. Those laws stood unchallenged as our best picture of the motion of the wandering stars and indeed the motion of all objects under gravity, from spinning galaxies to artillery shells, until Einstein's general theory of relativity came along in 1915.

This constantly shifting view of the position of the earth, the planets, and their motion through the heavens should serve as a lesson to anyone who is absolutely convinced that they know something. There are many things about the world that appear at first sight to be self-evidently true, and one of them is that we are standing still. Future observations can always surprise us, and they often do. Perhaps we should not be too surprised that nature sometimes appears counterintuitive to a tribe of observant, carbon-based ape descendants roaming around on the surface of a rocky world orbiting an unremarkable middle-aged star at the outer edge of the Milky Way galaxy. The theories of space and time we discuss in this book may well—in fact, probably will—turn out to be approximations to an as yet undiscovered deeper theory. Science is a discipline that celebrates uncertainty, and recognizing this is the key to its success.

Galileo Galilei was born twenty years after Copernicus proposed his sun-centered model of the universe, and he thought very deeply about the meaning of motion. His intuition would probably have been the same as ours: The earth feels to us as though it is standing still, although the evidence from the motion of the planets across the sky points very strongly to the fact that it is not. Galileo's great insight was to draw a profound conclusion from this seeming paradox. It feels like we are standing still, even though we know we are moving in orbit around the sun, because there is no way, not even in principle, of deciding what is standing still and what is moving. In other words, it only ever makes sense to speak of motion relative to something else. This is an incredibly important idea. It might seem obvious in some sense, but to fully appreciate its content requires some thought. It might seem obvious because, clearly, when you sit on the plane with your book, the book is not in motion relative to you. If you put it down on the table in front of you, it stays a fixed distance from you. And of course, from the point of view of someone on the ground, the book moves through the air along with the aircraft. The real content of Galileo's insight is that these statements are the only ones that can be made. And if all you can do is speak of how the book moves relative to you as you sit in your aircraft seat, or relative to the ground, or relative to the sun, or relative to the Milky Way, but always relative to something, then absolute motion is a redundant concept.

This rather provocative statement sounds superficially profound in the way that Zenlike utterances from fortune-tellers often do. In this case, however, it does turn out to be a great insight; Galileo deserves his reputation. To see why, let's say that we want to establish whether Aristotle's grid, which would allow us to judge whether something is in absolute motion, is useful from a scientific perspective. Useful in a scientific sense means that the idea has observable consequences. That means it has some kind of effect that can be detected by carrying out an experiment. By "experiment," we mean any measurement of anything at all; the swing of a pendulum, the color of light emitted by a burning candle flame, or the collisions of subatomic particles in the Large Hadron Collider at CERN (we'll come back to this experiment later on). If there are no observable consequences of an idea, then the idea is not necessary to understand the workings of the universe, although it might have some sort of chimerical value in making us feel better.

This is a very powerful way of sorting out the wheat from the chaff in a world full of diverse ideas and opinions. In his china teapot analogy, the philosopher Bertrand Russell illustrates the futility of holding on to concepts that have no observable consequences. Russell asserts that he believes there is a small china teapot orbiting between Earth and Mars, which is too small to be discovered by the most powerful telescopes in existence. If a larger telescope is constructed and, after an exhaustive and time-consuming survey of the entire sky, finds no evidence of the teapot, Russell will claim that the teapot is slightly smaller than expected but still there. This is commonly known as "moving the goalposts." Although the teapot may never be observed, it is an "intolerable presumption," says Russell, on the part of the human race to doubt its existence. Indeed, the rest of the human race should respect his position, no matter how preposterous it appears. Russell's point is not to assert his right to be left alone to his personal delusions, but that devising a theory that cannot be proved or disproved by observation is pointless in the sense that it teaches you nothing, irrespective of how passionately you may believe in it. You can invent any object or idea you like, but if there is no way of observing it or its consequences, you haven't made a contribution to the scientific understanding of the universe. Likewise, the idea of absolute motion will mean something in a scientific context only if we can devise an experiment to detect it. For example, we could set up a physics laboratory in an aircraft and carry out high-precision measurements on every conceivable physical phenomenon, in a last valiant attempt to detect our movement. We could swing a pendulum and measure the time it takes to tick, we could conduct electrical experiments with batteries, electric generators, and motors, or we could watch nuclear reactions take place and make measurements on the emitted radiation. In principle, with a big enough aircraft, we could carry out pretty much any and every experiment that has ever been conducted in human history. The key point that underpins this entire book and forms one of the very cornerstones of modern physics is that, provided the aircraft is not accelerating or decelerating, none of these experiments will reveal that we are in motion. Even looking out the window doesn't tell us this, because it is equally correct to say that the ground is flying past beneath us at six hundred miles per hour and that we are standing still. The best we can do is to say, "we are stationary relative to the aircraft," or "we are moving relative to the ground." This is Galileo's principle of relativity; there is no such thing as absolute motion, because it cannot be detected experimentally. This probably won't come as much of a shock, because we really do know it already at an intuitive level. A good example is the experience of sitting on a stationary train as the train on the next platform slowly pulls out of the station; for a split second it feels like we are the ones doing the moving. We find it difficult to detect absolute motion because there is no such thing.

This may all seem rather philosophical, but in fact such musings lead to a profound conclusion about the nature of space itself, and they allow us to take the first step along the path to Einstein's theories of relativity. So what conclusion about space can be drawn from Galileo's reasoning? The conclusion is this: If it is in principle impossible to detect absolute motion, it follows that there is no value in the concept of a special grid that defines "at rest," and therefore no value in the concept of absolute space.

This is important, so let us investigate it in more detail. We have already established that if it were possible to define a special Aristotelian grid covering the whole universe, then motion relative to that grid could be defined as absolute. We have also argued that since it is not possible to design an experiment that can tell us whether we are in motion, we should jettison the idea of that grid, simply because we can never work out to what it should be fixed. But how then should we define the absolute position of an object? In other words, where are we in the universe? Without the notion of Aristotle's special grid, these questions have no scientific meaning. All we can speak of are the relative positions of objects. There is therefore no way of specifying absolute positions in space, and that is what we mean when we assert that the notion of absolute space itself has no meaning. Thinking of the universe as a giant box, within which things move around, is a concept that is not required by experiment. We can't overemphasize how important this piece of reasoning is. The great physicist Richard Feynman once said that no matter how beautiful your theory, no matter how clever you are or what your name is, if it disagrees with experiment, it's wrong. In this statement is the key to science. Turning this statement around, if a concept is not testable by experiment, then we have no way of telling whether it's right or wrong, and it simply doesn't matter either way. Of course, that means we could still assume that an idea holds true, even if it isn't testable, but the danger is that in so doing we run the risk of hindering future progress because we are holding on to an unnecessary prejudice. So, without any possible means to identify a special grid, we are freed from the notion of absolute space, just as we have been freed from the concept of absolute motion. So what?! Well, being freed from the millstone of absolute space played a crucial role in allowing Einstein to develop his theory of space and time, but this will have to wait until the next chapter. For now, we have established our freedom, but we haven't acted as liberated scientists just yet. To whet the appetite, let us merely state that if there is no absolute space, then there is no reason why two observers should necessarily agree on the size of an object. That really should strike you as bizarre—surely if a ball has a diameter of 4 centimeters that is the end of the matter, but without absolute space it need not be.

So far we have discussed in some detail the connection between motion and space. What, then, of time? Motion is expressed as speed, and speed can be measured in miles per hour—that is, the distance traveled through space in a particular interval of time. In this way, the notion of time has in fact already entered into our thinking. Is there anything to be said of time? Is there some experiment we can do to prove that time is absolute, or should we also jettison this even more deeply held concept? Although Galileo dispensed with the notion of absolute space, his reasoning has nothing at all to teach us about absolute time. Time is immutable, according to Galileo. Immutable time means that it is possible to imagine perfect little clocks, all synchronized to show the same time, ticking away at every point in the universe. One clock could be on a plane, one on the ground, one (a tough one) at the surface of the sun, and one in orbit around a distant galaxy, and providing they are perfect timekeepers, they will read the same time as each other now and forever. Astonishingly, this seemingly obvious assumption turns out to be in direct contradiction with Galileo's statement that no experiment can tell us whether we are in absolute motion. Unbelievable as it may seem, the experimental evidence that finally destroyed the notion of absolute time emerged from the type of experiments many of us remember from school physics classes: batteries, wires, motors, and generators. To address the notion of absolute time, we must first take a detour into the nineteenth century, the golden age of discovery for electricity and magnetism.

2

The Speed of Light

Michael Faraday, the son of a Yorkshire blacksmith, was born in south London in 1791. He was self-educated, leaving school at fourteen to become an apprentice bookbinder. He engineered his own lucky break into the world of professional science after attending a lecture in London by the Cornish scientist Sir Humphry Davy in 1811. Faraday sent the notes he had taken at the lecture to Davy, who was so impressed by Faraday's diligent transcription that he appointed him his scientific assistant. Faraday went on to become a giant of nineteenth-century science, widely acknowledged to have been one of the greatest experimental physicists of all time. Davy is quoted as saying that Faraday was his greatest scientific discovery.

As twenty-first-century scientists, it is easy to look back at the early nineteenth century with envious eyes. Faraday didn't need to collaborate with 10,000 other scientists and engineers at CERN or launch a double-decker-bus-sized space telescope into high-earth orbit to make profound discoveries. Faraday's "CERN" fitted comfortably onto his bench, and yet he was able to make observations that led directly to the destruction of the notion of absolute time. The scale of science has certainly changed over the centuries, in part because those areas of nature that do not require technologically advanced apparatus to observe them have already been studied in exquisite detail. That's not to say there aren't examples in science today where simple experiments produce important results, just that to push back the frontiers across the board generally requires complicated machines. In early Victorian London, Faraday needed nothing more exotic or expensive than coils of wire, magnets, and a compass to provide the first experimental evidence that time is not what it seems. He gathered this evidence by doing what scientists like to do best. He set up all the paraphernalia associated with the newly discovered electricity, played around, and watched carefully. You can almost smell the darkly varnished bench mottled with shadows of coiled wire shifting in the gaslight, because even though Davy himself had dazzled audiences with demonstrations of electric lights in 1802 at the Royal Institution, the world had to wait until much later in the century for Thomas Edison to perfect a useable electric lightbulb. In the early 1800s, electricity was physics and engineering at the frontier of knowledge.

Faraday discovered that if you push a magnet through a coil of wire, an electric current flows in the wire while the magnet is moving. He also observed that if you send a pulse of electric current along a wire, a nearby compass needle is deflected in time with the pulse. A compass is nothing more than a magnet detector; when no electricity is pulsing through the wire, it will line up with the direction of the earth's magnetic field and point toward the North Pole. The pulse of electricity must therefore be creating a magnetic field like the earth's, although more powerful since the compass needle is wrenched away from magnetic north for a brief instant as the pulse moves by. Faraday realized that he was observing some kind of deep connection between magnetism and electricity, two phenomena that at first sight seem to be completely unrelated. What does the electric current that flows through a lightbulb when you flick a switch on your living room wall have to do with the force that sticks little magnetic letters to your fridge door? The connection is certainly not obvious, and yet Faraday had found by careful observation of nature that electric currents make magnetic fields, and moving magnets generate electric currents. These two simple phenomena, which now go by the name of electromagnetic induction, are the basis for generating electricity in all of the world's power stations and all of the electric motors we use every day, from the pump in your fridge to the "eject" mechanism in your DVD player. Faraday's contribution to the growth of the industrial world is incalculable.

Advances in fundamental physics rarely come from experiments alone, however. Faraday wanted to understand the underlying mechanism behind his observations. How could it be, he asked, that a magnet not physically connected to a wire can nevertheless cause an electric current to flow? And how can a pulse of electric current wrench a compass needle away from magnetic north? Some kind of influence must pass through the empty space between magnet, wire, and compass; the coil of wire must feel the magnet passing through it, and the compass needle must feel the current. This influence is now known as the electromagnetic field. We've already used the word "field" in the context of the earth's magnetic field, because the word is in everyday usage and you probably didn't notice it. In fact, fields are one of the more abstract concepts in physics. They are also one of the most necessary and fruitful for developing a deeper understanding. The equations that best describe the behavior of the billions of subatomic particles that make up the book you are now reading, the hand with which you are holding the book in front of your eyes, and indeed your eyes, are field equations. Faraday visualized his fields as a series of lines, which he called flux lines, emanating from magnets and current-carrying wires. If you have ever placed a magnet beneath a piece of paper sprinkled with iron filings, then you will have seen these lines for yourself. A simple example of an everyday quantity that can be represented by a field is the air temperature in your room. Near the radiator, the air will be hotter. Near the window, it will be cooler. You could imagine measuring the temperature at every point in the room and writing down this vast array of numbers in a table. The table is then a representation of the temperature field in your room. In the case of the magnetic field, you could imagine noting the deflection of a little compass needle at every point, and in that way you could form a representation of the magnetic field in the room. A subatomic-particle field is even more abstract. Its value at a point in space tells you the chance that the particle will be found at that point if you look for it. We will encounter these fields again in Chapter 7.

Why, you might legitimately ask, should we bother to introduce this rather abstract notion of a field? Why not stick to the things we can measure: the electric current and the compass needle deflections? Faraday found the idea attractive because he was at heart a practical man, a trait he shared with many of the great experimental scientists and engineers of the Industrial Revolution. His instinct was to create a mechanical picture of the connection between moving magnets and coils of wire, and for him the fields bridged the space between them to forge the physical connection his experiments told him must be present. There is, however, a deeper reason why the fields are necessary, and indeed why modern physicists see the fields as being every bit as real as the electric current and compass deflections. The key to this deeper understanding of nature lies within the work of Scottish physicist James Clerk Maxwell. In 1931, on the centenary of Maxwell's birth, Einstein described Maxwell's work on the theory of electromagnetism as "the most profound and the most fruitful that physics has experienced since the time of Newton." In 1864, three years before Faraday's death, Maxwell succeeded in writing down a set of equations that described all of the electric and magnetic phenomena Faraday and many others had meticulously observed and documented during the first half of the nineteenth century.

Equations are the most powerful of tools available to physicists in their quest to understand nature. They also are often among the scariest things most people meet during their school years, and we feel it necessary to say a few words to the apprehensive reader before we continue. Of course, we know that not everyone will feel that way about mathematics, and we ask for a degree of patience from more confident readers and hope they won't feel too patronized. At the simplest level, an equation allows you to predict the results of an experiment without actually having to conduct it. A very simple example, which we will use later in the book to prove all sorts of incredible results about the nature of time and space, is Pythagoras' famous theorem relating the lengths of the sides of a right-angled triangle. Pythagoras states that "the square of the hypotenuse is equal to the sum of the squares of the other two sides." In mathematical symbols, we can write Pythagoras' theorem as $x^2 + y^2 = z^2$, where z is the length of the hypotenuse, which is the longest side of the right-angled triangle, and x and y are the lengths of the other two sides. Figure 1 illustrates what is going on. The symbols x, y, and z are understood to be placeholders for the actual lengths of the sides and x^2 is mathematical notation for x multiplied by x. For example, $3^2 = 9$, $7^2 = 49$ and so on. There is nothing special about using x, y, and z; we could use any symbol we like as a placeholder. Perhaps Pythagoras' theorem looks more friendly if we write it as $☾^2 + ↟^2 = ☺^2$. This time the smiley-face symbol represents the length of the hypotenuse. Here is an example using the theorem: If the two shorter sides of the triangle are 3 centimeters (cm) and 4 centimeters long, then the theorem tells us that the length of the hypotenuse is equal to 5 centimeters, since $3^2 + 4^2 = 5^2$. Of course, the numbers don't have to be whole numbers. Measuring the lengths of the sides of a triangle with a ruler is an experiment, albeit a rather dull one. Pythagoras saved us the trouble by writing down his equation, which allows us to simply calculate the length of the third side of a triangle given the other two. The key thing to appreciate is that for a physicist, equations express relationships between "things" and they are a way to make precise statements about the real world.

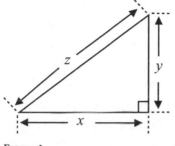

Maxwell's equations are mathematically rather more complicated, but in essence they do exactly the same kind of job. They can, for example, tell you in which direction a compass needle will be deflected if you send a pulse of electric current through a wire without having to look at the compass. The wonderful thing about equations, however, is that they can also reveal deep connections between quantities that are not immediately apparent from the results of experiments, and in doing so can lead to a much deeper and more profound understanding of nature. This turns out to be emphatically true of Maxwell's equations. Central to Maxwell's mathematical description of electrical and magnetic phenomena are the abstract electric and magnetic fields Faraday first pictured. Maxwell wrote down his equations in the language of fields because he had no choice. It was the only way of bringing together the vast range of electric and magnetic phenomena observed by Faraday and his colleagues into a single unified set of equations. Just as Pythagoras' equation expresses a relationship between the lengths of the sides of a triangle, Maxwell's equations express relationships between electric charges and currents and the electric and magnetic fields they create. Maxwell's genius was to invite the fields to emerge from the shadows and take center stage. If, for example, you asked Maxwell why a battery causes a current to flow in a wire, he might say, "because the battery causes there to be an electric field in the wire, and the field makes the current flow." Or if you asked him why a compass needle deflects near a magnet, he might say, "because there is a magnetic field around the magnet, and this causes the compass needle to move." If you asked him why a moving magnet causes a current to flow inside a coiled wire, he might answer that there is a changing magnetic field inside the coiled wire that causes an electric field to appear in the wire, and this electric field causes a current to flow. In each of these very different phenomena, the description always comes back to the presence of electric and magnetic fields, and the interaction of the fields with each other. Achieving a simpler and more satisfying view of many diverse and at first sight unrelated phenomena through the introduction of a new unifying concept is a common occurrence in physics. Indeed, it could be seen as the reason for the success of science as a whole. In Maxwell's case, it led to a simple and unified picture of all observed electric and magnetic phenomena that worked beautifully in the sense that it allowed for the outcome of any and all of the pioneering benchtop experiments of Faraday and his colleagues to be predicted and understood. This was a remarkable achievement in itself, but something even more remarkable happened during the process of deriving the correct equations. Maxwell was forced to add an extra piece into his equations that was not mandated by the experiments. From Maxwell's point of view, it was necessary purely to make his equations mathematically consistent. Contained in this last sentence is one of the deepest and in some ways most mysterious insights into the workings of modern science. Physical objects out there in the real world behave in predictable ways, using little more than the same basic laws of mathematics that Pythagoras probably knew about when he set about to calculate the properties of triangles. This is an empirical fact and can in no sense be said to be obvious. In 1960, the Nobel Prize–winning theoretical physicist Eugene Wigner wrote a famous essay titled "The Unreasonable Effectiveness of Mathematics in the Natural Sciences," in which he stated that "it is not at all natural that laws of Nature exist, much less that man is able to discover them." Our experience teaches us that there are indeed laws of nature, regularities in the way things behave, and that these laws are best expressed using the language of mathematics. This raises the interesting possibility that mathematical consistency might be used to guide us, along with experimental observation, to the laws that describe physical reality, and this has proved to be the case time and again throughout the history of science. We will see this happen during the course of this book, and it is truly one of the wonderful mysteries of our universe that it should be so.

To return to our story, in his quest for mathematical consistency, Maxwell added the extra piece, known as the displacement current, to the equation describing Faraday's experimental observations of the deflection of compass needles produced by electric currents flowing in wires. The displacement current was not necessary to describe Faraday's observations, and the equations described the experimental data of the time with or without it. Initially unbeknownst to Maxwell, however, with this simple addition his beautiful equations did far more than describe the workings of electric motors. With the displacement current included, a deep relationship between the electric and magnetic fields emerges. Specifically, the new equations can be recast into a form known as wave equations, which not surprisingly describe the motion of waves. Equations that describe the propagation of sound through the air are wave equations, as are equations that describe the journey of ocean waves to the shore. Quite unexpectedly, Maxwell's mathematical description of Faraday's experiments with wires and magnets predicted the existence of some kind of traveling waves. But whereas ocean waves are disturbances traveling through water, and sound waves are made up of moving air molecules, Maxwell's waves comprise oscillating electric and magnetic fields.

What are these mysterious oscillating fields? Imagine an electric field beginning to grow because Faraday generates a pulse of electric current in a wire. We have already learned that as the pulse of electric current passes along the wire, a magnetic field is generated (remember that Faraday observed that a compass needle in the vicinity of the wire is deflected). In Maxwell's language, the changing electric field generates a changing magnetic field. Faraday also tells us that when we change a magnetic field by pushing a magnet through a coil of wire, an electric field is generated, which causes a current to flow in the wire. Maxwell would say that a changing magnetic field generates a changing electric field. Now imagine removing the currents and the magnets. We are left with just the fields themselves, swinging backward and forward as changes in one generate changes in the other. Maxwell's wave equations describe how these two fields are linked together, oscillating backward and forward. They also predict that these waves must move forward with a particular speed. Perhaps not surprisingly, this speed is fixed by the quantities Faraday measured. In the case of sound waves, the wave speed is approximately 330 meters per second, just a little bit faster than a passenger airplane. The speed of sound is fixed by the details of the interactions between the air molecules that carry the wave. It changes with varying atmospheric pressure and temperature, which in turn describe how closely the air molecules get to each other and how fast they bounce off each other. In the case of Maxwell's waves, the speed is predicted to be equal to the ratio of the strengths of the electric and magnetic fields, and this ratio can be measured very easily. The strength of the magnetic field can be determined by measuring the force between two magnets. The word "force" will crop up from time to time, and by it we mean the amount by which something is pushed or pulled. The amount of push/pull can be quantified and measured, and if we are trying to understand how the world works, it should come as little surprise that we will want to understand how forces originate. In an equally simple way, the electric field strength can be measured by charging up two objects and measuring the force between them. You may have inadvertently experienced that "charging up" process yourself. Perhaps you've walked around over a nylon carpet on a dry day and then received an electric shock when you tried to open a door with a metal handle. This unpleasant door-opening experience occurs because you have rubbed electrons, the fundamental particles of electricity, off the carpet and into the soles of your shoes. You have become electrically charged, and this means that an electric field exists between you and the door handle. Given the chance when you grab hold of the door handle, this field will cause an electric current to flow, just as Faraday found in his experiments.

By carrying out such simple experiments, scientists can measure the strengths of the electric and magnetic fields, and Maxwell's equations predict that the ratio of strengths gives the speed of the waves. What, then, is the answer? What did Faraday's benchtop measurements, coupled with Maxwell's mathematical genius, predict for the speed of the electromagnetic waves? This is one of many key moments in our story. It is a wonderful example of why physics is a beautiful, powerful, and profound subject: Maxwell's waves travel at 299,792,458 meters per second. Astonishingly, this is the speed of light—Maxwell had stumbled across an explanation of light itself. You see the world around you because Maxwell's electromagnetic field drives itself through the darkness and into your eyes, at a speed predictable using only a coil of wire and a magnet. Maxwell's equations are the crack in the door through which light enters our story in a way that is every bit as important as the discoveries of Einstein that they triggered. The existence in nature of this special speed, a single, unchanging, 299,792,458 meters per second, will lead us in the next chapter, just as it led Einstein, to jettison the notion of absolute time.

The attentive reader might notice a puzzle here, or at least some sloppy writing on our part. Given what we said in Chapter 1, it clearly makes no sense to quote a speed without specifying relative to what that speed is defined, and Maxwell's equations make no mention of this problem. The speed of the waves—that is, the speed of light—appears as a constant of nature, the relationship between the relative strengths of the electric and magnetic fields. Nowhere in this elegant mathematical structure is there a place for the speed of the source of the waves, or indeed the receiver. Maxwell and his contemporaries knew this, of course, but it didn't worry them unduly. This is because most, if not all, of the scientists of the time believed that all waves, including light, must travel in some kind of medium; there must be some "real stuff" that is doing the waving. They were practical folk in Faraday's mold, and to them things don't just wave on their own with no support. Water waves can exist only in the presence of water, and sound waves travel only in the presence of air or some other substance, but certainly not in a vacuum: "In space, no one can hear you scream."

So the prevailing view at the end of the nineteenth century was that light must travel through a medium, and this medium was known as the ether. The speed that appeared in Maxwell's equations then had a very natural interpretation as the speed of light relative to the ether. This is exactly analogous to the propagation of sound waves through air. If the air is at a fixed temperature and pressure, then sound will always travel at a constant speed, which depends only on the details of the interactions between the air molecules, and has nothing to do with the motion of the source of the waves.

The ether must be a strange kind of stuff, though. It must permeate all of space, since light travels across the voids between the sun and earth and the distant stars and galaxies. When you walk down the street, you must be moving through the ether, and the earth must be passing through the ether on its yearly journey around the sun. Everything that moves in the universe must make its way through the ether, which must offer little or no resistance to the motion of solid objects, including things as large as planets. For if the ether did offer resistance to the motion of solid objects, the earth would have been slowed down during each of its 5 billion solar orbits, just as a ball bearing slows down when dropped into a jar of molasses, and our Earth years would gradually change in length. The only reasonable assumption must be that the earth and all objects move through the ether unimpeded. You may think that this would make its discovery impossible, but the Victorian experimentalists were nothing if not ingenious, and in a series of wonderfully high-precision experiments beginning in 1881, Albert Michelson and Edward Morley set out to detect the apparently undetectable. The experiments were beautifully simple in conception. In Bertrand Russell's excellent book on relativity written in 1925, he likens the earth's motion through the ether to going for a circular walk on a windy day; at some point you will be walking against the wind, and at some point with it. In a similar fashion, since the earth is moving through the ether as it orbits the sun, and the earth and sun together are flying through the ether in their journey around the Milky Way, then at some point in the year the earth must be moving against the ether wind, and at other times with it. And even in the unlikely event that the solar system as a whole is at rest relative to the ether, the earth's motion will still generate an ether wind as it travels around the sun, just as you feel the wind on your face when you stick your head out of the window of a moving car on a perfectly still day.

Michelson and Morley set themselves the challenge of measuring the speed of light at different times of the year. They and everyone else firmly expected that the speed would change over the course of a year, albeit by a tiny amount, because the earth (and along with it their experiment) should be constantly changing its speed relative to the ether. Using a technique called interferometry, the experiments were exquisitely sensitive, and Michelson and Morley gradually refined the technique over six years before publishing their results in 1887. The result was un-equivocally negative. No difference in the speed of light in any direction and at any time of year was observed.

If the ether hypothesis is correct, this result is very hard to explain. Imagine, for example, that you decide to dive into a fast-flowing river and swim downstream. If you swim at 5 kilometers per hour through the water, and the river is flowing at 3 kilo-meters per hour, then relative to the bank you will be swimming along at 8 kilometers per hour. If you turn and swim back up-stream, then relative to the bank you will be swimming at 2 kilo-meters per hour. Michelson and Morley's experiment is entirely analogous: You, the swimmer, are the beam of light, the river is the ether through which the light is supposed to travel, and the riverbank is Michelson and Morley's experimental apparatus, sat at rest on the earth's surface. Now we can see why the Michelson-Morley result was such a surprise. It was as if you always travel at 5 kilometers per hour relative to the riverbank, irrespective of the river's speed of flow and the direction in which you decide to swim.

So Michelson and Morley failed to detect the presence of an ether flowing through their apparatus. Here is the next chal-lenge to our intuition: Given what we have seen so far, the bold thing to do might be to jettison the notion of the ether because its effects cannot be observed, just as we jettisoned the notion of absolute space in Chapter 1. As an aside, from a philosophical perspective the ether was always a rather ugly concept, since it would define a benchmark in the universe against which absolute motion could be defined in conflict with Galileo's principle of relativity. Historically, it seems likely that this was Einstein's personal view, because he appears to have been only vaguely aware of Michelson and Morley's experimental results when he took the bold step of abandoning the ether in formulating his special theory of relativity in 1905. It is certainly the case, however, that philosophical niceties are not a reliable guide to the workings of nature and, in the final analysis, the most valid reason to reject the ether is that the experimental results do not require it.*

While the rejection of the ether may be aesthetically pleasing and supported by the experimental data, if we choose to take this plunge then we are certainly left with a serious problem: Maxwell's equations make a very precise prediction for the speed of light but contain no information at all about relative to what that speed should be measured. Let us for a moment be bold, accept the equations at face value, and see where the intellectual journey leads. If we arrive at nonsense, then we can always backtrack and try another hypothesis, feeling satisfied that we have done some science. Maxwell's equations predict that light always moves with a velocity of 299,792,458 meters per second, and there is no place to insert the velocity of the source of the light or the velocity of the receiver. The equations really do seem to assert that the speed of light will always be

---

* There have been many attempts, since Michelson and Morley, to detect the ether and all have yielded null results.

measured to be the same, no matter how fast the source and the receiver of the light are moving relative to each other. It seems that Maxwell's equations are telling us that the speed of light is a constant of nature. This really is a bizarre assertion, so let us spend a little more time exploring its meaning.

Imagine that light is shining out from a flashlight. According to common sense, if we run fast enough we could in principle catch up with the front of the beam of light as it advances forward. Common sense might even suggest that we could jog alongside the front of the beam if we managed to run at the speed of light. But if we are to follow Maxwell's equations to the letter, then no matter how fast we run, the beam still recedes away from us at a speed of 299,792,458 meters per second. If it did not, the speed of light would be different for the person running compared to the person holding the flashlight, contradicting Michelson and Morley's experimental results and our assertion that the speed of light is a constant of nature, always the same number, irrespective of the motion of the source or the observer. We seem to have talked ourselves into a ridiculous position. Surely common sense would advise us to reject, or at least modify or reinterpret Maxwell's equations: Perhaps they are only approximately correct. Now, that doesn't sound like an unreasonable proposition, since the motion of any realistic experimental apparatus would cause only a tiny variation in the 300 million meters per second that appears in Maxwell's equations. So tiny indeed that perhaps it would have remained undetected in Faraday's experiments. The alternative is to accept the validity of Maxwell's equations and the bizarre proposition that we can never catch up with a beam of light. Not only is that idea an outrage to our common sense, but the next chapter will reveal that it also implies that we should reject the very notion of absolute time.

Breaking our attachment to absolute time is just as difficult to grasp today as it was to the nineteenth-century scientists. We have a strong intuition in favor of absolute space and time that is very hard to break, but we should be clear that intuition is all it is. Moreover, Newton's laws embrace these notions whole-heartedly and, even to this day, those laws underpin the work of many engineers. Back in the nineteenth century, Newton's laws seemed untouchable. While Faraday was laying bare the work-ings of electricity and magnetism at the Royal Institution, Isambard Kingdom Brunel was driving the Great Western Railway from London to Bristol. Brunel's iconic Clifton Suspension Bridge was completed in 1864, the same year that Maxwell achieved his magnificent synthesis of Faraday's work and un-covered the secret of light. The Brooklyn Bridge opened eight years later, and by 1889 the Eiffel Tower had risen above the Paris skyline. All of the magnificent achievements of the age of steam were designed and built using the concepts laid down by Newton. Newtonian mechanics was clearly far from being ab-stract mathematical musing. The symbols of its success were rising across the face of the globe in an ever-expanding cele-bration of humanity's mastery of the laws of nature. Imagine the consternation in the minds of the late nineteenth–century scientists when they were faced with Maxwell's equations and their implicit attack on the very foundations of the Newtonian worldview. Surely there could be only one winner. Surely Newton and the notion of absolute time would reign victorious. Nevertheless, the twentieth century dawned with the problem of the constant speed of light still casting dark clouds: Maxwell and Newton could not both be right. It took until 1905 and the work of a hitherto unknown physicist named Albert Einstein for it to be finally demonstrated that nature sides with Maxwell.

3

Special Relativity

In Chapter 1 we succeeded in establishing that the very intuitive Aristotelian view of space and time was laden with excess baggage. That is to say, we showed that there is simply no need to view space as the fixed, immutable, and absolute structure in which things happen. We also saw how Galileo appreciated the irrelevance of holding on to the notion of absolute space, while firmly maintaining the idea of a universal time. In the last chapter, we took a detour into the nineteenth-century physics of Faraday and Maxwell, where we learned that light is none other than a symbiosis of electric and magnetic fields surging forward in perfect agreement with Maxwell's beautiful equations. Where does all that leave us? If we are to dismiss the idea of absolute space, with what are we to replace it? And what does it mean when we allude to the breakdown of the notion of absolute time? The aim of this chapter is to provide answers to these questions.

Albert Einstein is undoubtedly the iconic figure of modern science. His white, unkempt hair and sockless demeanor provide the contemporary shorthand for "professor"; ask a child to draw a scientist and she might well produce something that looks like the old Einstein. The ideas in this book are, however, the ideas of a young man. At the turn of the twentieth century, when Einstein was thinking about the nature of space and time, he was in his early twenties, with a young wife and family. He did not have an academic post at a university or research establishment, although he discussed physics regularly with a small group of friends, often late into the night. An unfortunate consequence of Einstein's apparent isolation from the mainstream is the modern temptation to look upon him as a maverick who took on the scientific establishment and won; unfortunate because it provides inspiration to any number of crackpots who think they have single-handedly discovered a new theory of the universe and cannot understand why nobody will listen to them. In fact, Einstein was reasonably well connected to the scientific establishment, although it is true that he did not have an easy beginning to his academic career.

What is striking is his persistence in continuing to explore the important scientific problems of the day while being overlooked for university-level academic positions. On emerging from the Swiss Federal Institute of Technology (ETH) in Zürich at the age of twenty-one, having qualified as a specialized teacher in science and mathematics, he took a series of temporary teaching positions that allowed him the time to work on his doctoral thesis. During 1901, while teaching at a private school in Schaffhausen in northern Switzerland, he submitted his doctoral thesis to the University of Zürich, which was rejected. Following that setback, Einstein moved to Bern and famously began his career as a technical expert, third class, in the

Swiss patent office. The relative financial stability and freedom this afforded resulted in the most productive years of his life, and arguably the most productive years of any single scientist in history.

Most of this book deals with Einstein's work leading up to and encompassing his golden year of 1905, in which he first wrote down $E = mc^2$, was finally awarded his PhD, and completed a paper on the photoelectric effect, for which he eventually won the Nobel Prize. Remarkably, Einstein was still working at the patent office in 1906, where his reward for changing our view of the universe forever was to be promoted to technical expert, second class. He finally got a "proper" academic position in Bern in 1908. While one might be tempted to wonder what Einstein could have achieved if he had not been forced to relegate physics to a leisure pursuit during these years, he always looked back with immense fondness at his time in Bern. In his book *Subtle Is the Lord*, Einstein's biographer and friend, Abraham Pais, described Einstein's days at the patent office as "the closest he would ever come to paradise on earth," because he had the time to think about physics.

Einstein's inspiration on the road to $E = mc^2$ was the mathematical beauty of Maxwell's equations, which impressed him to such a degree that he decided to take seriously the prediction that the speed of light is a constant. Scientifically this doesn't sound like too controversial a step: Maxwell's equations were built on the foundation of Faraday's experiments, and who are we to argue with the consequences? All that stands in our way is a prejudice against the notion that something can move at the same speed regardless of how fast we chase after it. Imagine driving down a road at 40 miles per hour and suppose a car passes you traveling at a speed of 50 miles per hour. It seems to be pretty obvious that you see the second car pull away at a net speed of 10 miles per hour. Thinking of this as "obvious" is just the kind of prejudice that we have to resist if we are to follow Einstein and accept that light always streams away from you at the same speed regardless of how fast you are moving. Let us for now accept, as Einstein did, that our common sense might be misleading us, and see where a constant speed of light will lead.

At the heart of Einstein's theory of special relativity lie two proposals, which in the language of physics are termed axioms. An axiom is a proposition that is assumed to be true. Given the axioms, we can then proceed to work out the consequences for the real world, which we can check using experiments. The first part of this method is an old one, dating back to ancient Greece. Euclid most famously deployed it in his *Elements*, in which he developed the system of geometry still taught in schools to this day. Euclid constructed his geometry based on five axioms, which he took to be self-evident truths. As we shall see later, Euclid's geometry is in fact only one of many possible geometries: the geometry of a flat space, such as a tabletop. The geometry of the surface of the earth is not Euclidean and is defined by a different set of axioms. Another even more important example for us, as we shall soon learn, is the geometry of space and time. The second part, checking the consequences against nature, was not much used by the ancient Greeks. If it had been, then the world might well be a very different place today. This seemingly simple step was introduced to the world by Muslim scientists as early as the eleventh century and took hold in Europe much later, in the sixteenth and seventeenth centuries. With the anchor of experiment, science was finally able to make rapid progress, and with that came technological advancement and prosperity.

The first of Einstein's axioms is that Maxwell's equations hold true in the sense that light always travels through empty space at the same speed regardless of the motion of the source or the observer. The second axiom advocates that we are to follow Galileo in asserting that no experiment can ever be performed that is capable of identifying absolute motion. Armed only with these propositions, we can now proceed as good physicists should and explore the consequences. As ever in science, the ultimate test of Einstein's theory, derived from his two axioms, is its ability to predict and explain the results of experiments. Quoting Feynman more fully this time: "In general we look for a new law by the following process. First we guess it. Then we compute the consequences of the guess to see what would be implied if this law that we guessed is right. Then we compare the result of the computation to Nature, with experiment or experience, compare it directly with observation, to see if it works. If it disagrees with experiment it is wrong. In that simple statement is the key to science. It does not make any difference how beautiful your guess is. It does not make any difference how smart you are, who made the guess, or what his name is—if it disagrees with experiment it is wrong. That's all there is to it." It is a terrific quote from a lecture filmed in 1964, and we recommend looking it up on YouTube.

Therefore, our goal for the next few pages is to work out the consequences of Einstein's axioms. We will begin by using a technique that Einstein himself often favored: the thought experiment. Specifically, we want to explore the consequences of assuming that the speed of light remains constant for all observers, no matter how they are moving relative to each other. To do this, we are going to imagine a clumsy-looking clock called a light clock. The clock consists of two mirrors, between which a beam of light bounces back and forth. We can use this as a clock by counting each bounce of the light beam as one tick. For example, if the mirrors are 1 meter apart, then it takes light approximately 6.67 nanoseconds to complete one round trip.* You can check this number for yourself: The light has to travel 2 meters and does so at a speed of 299,792,458 meters every second. This would be a very high-precision clock, with around 150 million ticks corresponding to one heartbeat.

Now, imagine putting the light clock on a train that is whizzing along past someone standing on a station platform. The million-dollar question is: How fast does the clock on the train tick according to the person on the platform? Until Einstein, everybody assumed that it ticks at the same rate—one tick every 6.67 nanoseconds.

Figure 2 shows how one tick of the clock on the train looks according to the person standing on the platform. Because the train is moving, the light must travel farther in one tick, as determined from the platform. Put another way, the starting point of the light beam's journey is not in the same place as its end point according to the person on the platform, because the clock has moved during the tick. In order for the clock to tick at the

---

* A nanosecond is one thousandth of a microsecond, or 0.000000001 seconds.

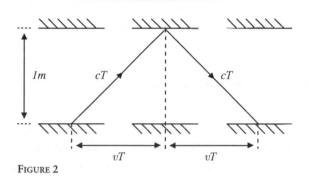

FIGURE 2

same rate as it does when it stands still the light must travel a little bit faster. Otherwise it could not complete its longer journey in 6.67 nanoseconds. This is exactly what happens in Newton's worldview, because the light is given a helping hand by the motion of the train. But—and this is the crucial step—applying Einstein's logic means that the light *cannot* speed up because the speed of light must be the same to everyone. This has the disturbing consequence that the moving clock must genuinely take longer to tick, simply because the light has farther to travel, from the perspective of the person on the platform. This thought experiment teaches us that if we are to assert that the speed of light is a constant of nature, as Maxwell seems to be trying to tell us, then it follows that time ticks at different rates depending on how we are moving relative to someone else. In other words, absolute time is not consistent with the notion of a universal light speed.

It is very important to emphasize that this conclusion is not specific to light clocks. There is no important difference between a light clock and a pendulum clock, which works by "bouncing" the pendulum between two places once every second. Or for that matter an atomic clock, which counts the number of peaks and troughs of a light wave emitted from an atom to generate the ticks. Even the rate of decay of the cells in your body could be used as little clocks, and the conclusions would be the same because all these devices measure the passing of time. The light clock is in fact a bit of an old chestnut in the teaching of Einstein's theory and provokes no end of confused discussion because it is such an unfamiliar clock. It can be tempting to attribute the weird conclusion we have just reached to this lack of familiarity, rather than to recognize it as an insight into the nature of time itself. To do so would be to make a bad mistake— our sole reason for picking a light clock rather than any other type of clock is that we can exploit Einstein's bizarre demand that light should travel at the same speed for everyone to draw our conclusions. Any conclusion that we draw from thinking about the light clock must also apply to any other kind of clock, for the following reason. Imagine that we seal ourselves into a box with a light clock and a pendulum clock and set them ticking away in sync. If they are very accurate clocks, they will stay in sync and tell the same time forever. Now, let's put the box onto the moving train. According to Einstein's second axiom, we should not be able to tell whether we are moving. But if the light clock behaved differently than the pendulum clock, they would drift out of sync and we could say for certain from inside our sealed box that we were moving.* So a pendulum clock

---

* The sealed box is just to stop us from being distracted by the idea that we could look out of the window of the train to determine whether we are moving. Of course that is irrelevant; looking out of the window only ascertains that we are moving relative to the ground outside.

and a light clock must count time in exactly the same way and that means that if the moving light clock is running slow as determined by the person on the platform, then so too must all other moving clocks run slow. This isn't some kind of optical illusion: The passage of time is slowed down on the moving train according to someone on the platform.

The upshot is that we must either cling to the comforting notion of absolute time and ditch Maxwell's equations, or ditch absolute time in favor of Maxwell and Einstein. How should we check which is the correct thing to do? We must find an experiment in which we should, if Einstein is right, observe time actually slowing down for moving objects.

To design such an experiment, first we need to work out how fast something should move in order to reveal the proposed effect. It should be quite clear that moving at 70 mph down the highway in a car does not cause time to slow down very much, because we don't come home after a trip to the store to find that our children have grown older than us while we were away. Silly as this seems, taking Einstein at face value means that this is exactly what does happen, and we would certainly notice the difference if only we could travel fast enough. So what constitutes fast enough? From the viewpoint of the person on the station platform, the light travels along the two sides of the triangle shown in the diagram. Einstein's argument is that because this is a greater distance for the light to travel than if the clock were standing still, time will pass more slowly because the tick takes longer. All we have to do now is to calculate how much longer (for a given train speed) and we have the answer. We can do this with a little help from Pythagoras.

If you do not want to follow the maths you can skip over the next paragraph, but then you will have to take our word for it that the numbers all work out. That goes for any other maths we might bump into as the book progresses. It is always an option to skip past it and not worry—the mathematics helps provide a deeper appreciation of the physics but it isn't absolutely necessary to follow the flow of the book. Our hope is that you will have a go with the maths even if you have no prior experience at all. We have tried to keep things accessible. Perhaps the best way to approach the maths is not to worry about it. The logic puzzles that appear in the daily newspapers are much harder to tackle than anything we will do in this book. That said, here comes one of the trickier bits of maths in the book, but the result is worth the effort.

Take a look at Figure 2 again and suppose that the time taken for half of one tick of the clock on the train as measured by the person standing on the platform is equal to T. It is the time taken for the light to travel from the bottom mirror to the top mirror. Our goal is to figure out what T actually is and double it to get the time for one tick of the clock according to the person on the platform. If we did know T, then we could figure out that the length of the longest side of the triangle (the hypotenuse) is equal to cT, i.e., the speed of light (c) multiplied by the time taken for light to travel from the bottom mirror to the top mirror (T). Remember, the distance something travels is obtained by multiplying its speed by the time of the journey. For example, the distance a car travels in one hour at 60 miles per hour is 60 x 1 = 60 miles. It is not hard to work out the result for a two-hour journey. All we are doing here is invoking the formula "distance = speed x time." Knowing T, we could also figure out how far the clock moves in half of one tick. If the train is moving at a speed, v, then the clock moves a distance vT each half-tick. Again we did nothing except use "distance = speed x time." This distance is the length of the base of a right-angled triangle and because we know the length of the longest side, we can go ahead and figure out the distance between the two mirrors using Pythagoras' theorem. But we know what that distance actually is already—it is 1 meter. So Pythagoras' theorem tells us that $(cT)^2 = 1^2 + (vT)^2$. Note the use of parentheses: In mathematics they are used to indicate which operations to carry out first. In this case $(vT)^2$ means "multiply v by T and then square up the answer." That's all there is to it.

We are nearly done now. We know c, the speed of light, and let's presume to know the speed of the train, v. Then we can use this equation to figure out T. The crudest way to do it would be to guess a value of T and see if it solves the equation. More often than not the guess will be wrong and we'll need to try another guess. After a while we might hone in on the right answer. Fortunately, we can avoid that tedious process because the equation can be "solved." The answer is $T^2 = 1/(c^2 - v^2)$, which means, "first work out $c^2 - v^2$ and then divide 1 by that number." The forward slash is the symbol we will use to denote "divide by." So $1/2 = 0.5$ and a/b means "a divided by b," etc. If you know a bit of maths, then you'll probably feel a little bored by now. If not, then you might wonder how we arrived at $T^2 = 1/(c^2 - v^2)$. Well, this isn't a book on maths, and you'll just have to trust that we got it right—you can always convince yourself that we got it right by putting some numbers in. Actually, we

have the result for T^2, which means "T multiplied by T." We get
T by taking the square root. Mathematically, the square root of
a number is such that when multiplied by itself we regain the
original number; for example, the square root of 9 is 3 and the
square root of 7 is close to 2.646. There is a button on most cal-
culators that computes the square root for you. It is usually denoted by the symbol "$\sqrt{}$" and one would normally write things
like $3 = \sqrt{9}$. As you can see, the square root is the opposite of
squaring, $4^2 = 16$ and $\sqrt{16} = 4$.

Returning to the task at hand, we can now write the time
taken for one tick of the clock as determined by someone on
the platform: It is the time for light to travel up to the top mirror and back down again—that is $2T$. Taking the square root of
our equation above for T^2, and multiplying by 2, we find that
$2T = 2/\sqrt{c^2 - v^2}$. This equation allows us to work out the time
taken for one tick as measured by the person on the platform,
knowing the speed of the train, the speed of light, and the distance between the two mirrors (1 meter). But the time for one
tick according to someone sitting on the train next to the clock
is simply equal to $2/c$, because for them the light simply travels
2 meters at a speed c (distance = speed x time, so time = distance / speed). Taking the ratio of these two time intervals tells
us by how much the clock on the train is running slow, as measured by someone on the platform; it is running slow by a factor of $c/\sqrt{c^2 - v^2}$, which can also be written, with a little more
mathematical rearranging, as $1/\sqrt{1 - v^2/c^2}$. This is a very important quantity in relativity theory, and it is usually represented
by the Greek letter γ, pronounced "gamma." Notice that γ is always larger than 1 as long as the clock is flying along at less than

the speed of light, because v/c will be smaller than 1. When v is very small compared to the speed of light (i.e., for most ordinary speeds, since in units more familiar to motorists the speed of light is 671 million miles per hour), γ is very close to 1 indeed. Only when v becomes a significant fraction of the speed of light does γ start to deviate appreciably from 1.

Now we are done with the mathematics—we have succeeded in figuring out by exactly how much time slows down on the train as determined by someone on the platform. Let's put some numbers in to get a feel for things. If the train is moving at 300 kilometers per hour, then you can check that v^2/c^2 is a very tiny number: 0.000000000000077. To get the "time stretching" factor γ we need $1/\sqrt{1-0.000000000000077}$ = 1.000000000000039. As expected, it is a tiny effect: Traveling for 100 years on the train would only extend your lifetime by a matter of 0.0000000000039 years according to your friend on the platform, which is slightly above one-tenth of a millisecond. The effect would not be so tiny if the train could whiz along at 90 percent of the speed of light, however. The time stretching factor would then be bigger than 2, which means that the moving clock would tick at less than half the rate of the station clock according to someone sitting on the platform. This is Einstein's prediction and, like all good scientists, we have to test it experimentally if we are to believe it. It certainly seems a little unbelievable at this point.

Before we discuss an experiment that settles the argument, let us pause to reflect upon the result that we just uncovered. Let's look once again at the thought experiment from the point of view of a passenger on the train sitting beside the clock. For the passenger, the clock is not moving and the light simply bounces up and down, just as it would have for a person sitting with the same clock in a café in the station. The passenger must see the clock tick once every 6.67 nanoseconds and 150 million times for every heartbeat, because she is perfectly correct in deciding that the clock is not moving relative to herself, in the spirit of Galileo. Meanwhile, the person on the platform says that the clock on the train took a little longer than 6.67 nanoseconds to perform one tick. After 150 million ticks of the moving clock, his heart will therefore have made slightly more than one heartbeat. This is astonishing: According to the person on the platform, he is aging faster than the passenger sitting on the train.

As we have just seen, the effect is a tiny one for real trains, which don't travel anywhere near as fast as the speed of light, but it is real nonetheless. In an imaginary world where the train whizzes along a very long track at close to the speed of light, the effect gets magnified and there would be no doubt about it: The person on the platform would age quicker from his perspective.

In real experiments, if we are to test this breakdown in absolute time, then we need to find a way to investigate objects that can move close to the speed of light, for only then will the time-stretching factor γ become measurably larger than 1. Ideally we'd also like to study an object that has a lifetime, that is to say that it dies. We could then look to see if we could prolong the lifetime of the object simply by making it move fast.

Fortunately for scientists, such objects do exist; in fact, scientists themselves are built out of them. Elementary particles are tiny subatomic objects that by virtue of their smallness are easy to accelerate to vast speeds. They are referred to as elementary because, as far as we can tell with our current technology, they are the smallest building blocks of everything in the universe. We will have much more to say about elementary particles later in the book. For now, we would like to describe just two: the electron and the muon.

The electron is a particle to which we are all indebted, because we are built out of them. It is also the particle that flows through electric wires to light our bulbs and heat our ovens; the electron is the particle of electricity. The muon is identical to the electron in every way, except it is heavier. Why nature should have chosen to give us a copy of the electron that appears to be redundant if all you want to do is to build planets and people, is not something physicists really understand. Whatever the reason for the existence of the muon, it is very useful indeed to scientists wishing to test Einstein's theory of relativity because it has a short lifetime and it is very small and easy to accelerate to very high speeds. As far as we can tell, electrons live forever, whereas a muon placed at rest beside you would live for something like 2.2 microseconds (a microsecond is one-millionth of a second). When a muon dies, it nearly always turns into an electron and another pair of subatomic particles called neutrinos, but that is extra information that we don't need. All we need here is that the muon does die. The Alternating Gradient Synchrotron (AGS) facility at Brookhaven National Laboratory on Long Island, New York, provides a very nice test of Einstein's theory. In the late 1990s, the scientists at Brookhaven built a machine that produced beams of muons circulating around a 14-meter-diameter ring at a speed of 99.94 percent of the speed of light. If

muons live for only 2.2 microseconds when they are speeding
around the ring, then they would manage only 15 laps of the ring
before they died.* In reality, they managed more like 400 laps,
which means their lifetime is extended by a factor of 29 to just
over 60 microseconds. This is an experimental fact. Einstein appears to be on the right track, but just how accurate is he?

Here is where the mathematics we did earlier in this chapter
becomes very valuable. We have made a precise prediction for
the amount by which a little clock traveling at speed actually
slows down relative to a clock standing still. We can therefore
use our equation to predict by how much time should slow down
when traveling at 99.94 percent of the speed of light, and therefore by how much a muon's lifetime should be extended. Einstein predicts that the muons in Brookhaven should have their
time stretched by a factor of $\gamma = 1/\sqrt{1 - v^2/c^2}$ with $v/c =$
0.9994. If you have a calculator handy, then type the numbers in
and see what happens. Einstein's formula gives 29, exactly as the
Brookhaven experimenters found.

It's worth taking a brief pause here to ponder what has happened. Using only Pythagoras' theorem and Einstein's assumption about the speed of light being the same for everyone, we
derived a mathematical formula that allowed us to predict the
lengthening of the lifetime of a subatomic particle called a
muon when that muon is accelerated around a particle accelerator in Brookhaven to 99.94 percent of the speed of light. Our
prediction was that it should live 29 times longer than a muon

---

* You can check this for yourself once you know that the circumference of
a circle is equal to pi multiplied by the diameter, where pi equals approximately
3.142.

standing still, and this prediction agrees exactly with what was seen by the scientists at Brookhaven. The more you think about this, the more wonderful it is. Welcome to the world of physics! Of course, Einstein's theory was already well established in the late 1990s. The scientists at Brookhaven were interested in studying other properties of their muons—the life-enhancing effects of Einstein's theory provided a bonus, which meant that they got to observe them for longer.

We must therefore conclude, because experiment tells us so, that time is malleable. Its rate of passage varies from person to person (or muon to muon) depending upon how they move about.

As if this rather unsettling behavior of time isn't enough, something else is lurking, and the alert reader may have spotted it. Think back to those muons whizzing around the AGS. Let's put a little finish line in the ring and count how many times the muons cross it as they circulate before they die. For the person watching the muons, they cross 400 times because their life-times have been extended. How many times would you cross the finish line if you could speed around the ring with the muons? It has to be 400 as well, of course; otherwise the world would make no sense at all. The problem is that according to your watch, as you fly around the ring with the muons, they live for only 2.2 microseconds, because the muon is standing still relative to you and muons live for 2.2 microseconds when they stand still. Nevertheless, you and the muon must still manage to make 400 or so laps of the ring before the muon finally expires. What has happened? Four hundred laps in 2.2 microseconds doesn't seem possible. Fortunately, there is a way out of this dilemma. The circumference of the ring could be reduced from the viewpoint of the muon. To be entirely consistent, the length of the ring, as determined by you and the muon, must shrink by exactly the same amount that the muon's lifetime increases. So space must be malleable too! As with the stretching of time, this is a real effect. Real objects do shrink when they move. As a bizarre example, imagine a 4-meter-long car trying to fit into a 3.9-meter-long garage. Einstein predicts that if the car is traveling faster than 22 percent of the speed of light, then it will just about squeeze into the garage, at least for a split second before it crashes through the back wall. Again, if you have been following the maths, then you can check that 22 percent is the right number. Any faster and the car shrinks to below 3.9 meters; any slower and it doesn't shrink enough.

The discovery that the passage of time can be slowed down and distances can be shrunk is strange enough when applied to the realm of subatomic particles, but Einstein's reasoning applies equally well to things the size of humans. One day we may even come to rely on this strange behavior for our survival. Imagine living on the earth in the far future. In a few billion years' time, the sun will no longer be a stable provider of life-sustaining illumination to our world, but a seething, unstable monster of a star that may well engulf our planet as it swells in its final reddening death throes. If we have not become extinct for some other reason by then, it will be necessary for humans to escape our ancestral home and journey to the stars. The Milky Way, our local spiral island of a hundred billion suns, is 100,000 light-years across. This means that light takes 100,000 years to journey across it, as determined by someone on Earth. Hopefully, the need for the last caveat is clear given all that we have been saying. It might seem that humanity's possible destinations within the Milky Way will be forever restricted to a tiny portion of the stars very close to our home (on astronomical scales) because we could hardly be expected to undertake a journey to distant corners of the galaxy that would take light itself 100,000 years to reach. But here is where Einstein comes to the rescue. If we could build a spaceship that could whisk us into space at speeds very close to light speed, then the distances to the stars would shrink, and the amount of shrinking would increase the closer to light speed we could travel. If we managed to travel at 99.99999999 percent of light speed, then we could travel out of the Milky Way and all the way to the neighboring Andromeda galaxy, almost 3 million light-years away, in a mere fifty years. Admittedly, that looks like a tall order and indeed it is. The big obstacle is figuring out how to power a spaceship so that it could get up to such high speeds, but the point remains: With the warping of space and time, travel to distant parts of the universe becomes imaginable in a way it never was before. If you were part of humanity's first Andromeda expedition, arriving in a new galaxy after a fifty-year journey, your children born in space might wish to return to their home world and gaze upon the earth with their own eyes for the first time. For them, the Blue Planet would be nothing more than a bedtime space story. Turning the spaceship around, and traveling back to Earth for fifty years, the entire journey to Andromeda and back would have taken one hundred years. By the time they arrived back in Earth orbit, however, a shocking 6 million years would have passed by for the inhabitants of the earth. Would their progenitor civilization have even survived? Einstein has opened our eyes to a weird and wonderful world.

4

Spacetime

In the previous chapters we followed the historical road to relativity, and in fact our reasoning was not too far from what Einstein originally presented. We have been forced to accept that space is not the great stage upon which the events of our lives are played out. Likewise, time is not something universal and absolute. Instead we moved toward a picture of space and time that is much more malleable and subjective. The great clock in the sky, and in some sense the sky itself, has been banished. It might feel to us like the world is a box within which we go about our business, because that picture allows us to make sense of it quickly and efficiently. The ability to map the movement of things against an imaginary grid is what we might call spatial awareness, and it is clearly important if you are to avoid predators, catch food, and survive in a dangerous and challenging environment. But there is no reason why this model, buried deep within our brains and reinforced over millions of years by natural selection, should be anything other than a model. If a way of thinking about the world confers a survival advantage, then that way of thinking will become ubiquitous.

The scientific correctness of it is irrelevant. The important thing is that, because we chose to accept the results of experiments conducted on Faraday's mottled benchtop and the explanations codified by Maxwell, we have acted like scientists and rejected the comfortable model of space and time that allowed our ancestors to survive and prosper on the ancient plains of Africa. This model has been embedded and reinforced deep within our psyche by our experiences over many millions of years, and discarding it may well be disorientating. That dizzying feeling of confusion, if (hopefully) followed by an epiphany of clarity, is the joy of science. If the reader is feeling the former, we hope to deliver the latter by the end of the book.

This is not a history book. Our aim is to describe space and time in the most enlightening way we can, and it is our view that the historical road to relativity does not necessarily provide the best path to enlightenment. From a modern perspective, over a century after Einstein's revolution, we have learned that there is a deeper and more satisfying way to think about space and time. Rather than dig any deeper into the old-fashioned textbook view, we are going to start again from a blank canvas. In so doing we will come to understand what Minkowski meant when he said that space and time must be merged together into a single entity. Once we have developed a more elegant picture, we will be well placed to achieve our principal goal—we shall be able to derive $E = mc^2$.

Here is the starting point. Einstein's theories can be constructed almost entirely using the language of geometry. That is, you don't need much algebra, just pictures and concepts. At the heart of the matter, there lie only three concepts: invariance, causality, and distance. Unless you are a physicist, two of these will probably be unfamiliar words, and the third familiar but, as we shall see, subtle.

Invariance is a concept that lies at the core of modern physics. Glance up from this book now and look out at the world. Now turn around and look in the opposite direction. Your room will look different from different vantage points, of course, but the laws of nature are the same. It doesn't matter whether you are pointing north, south, east, or west, gravity still has the same strength and still keeps your feet on the ground. Your TV still works when you spin it around, and your car still starts whether you've left it in London, Los Angeles, or Tokyo. These are all examples of invariance in nature. When put like this, invariance seems like little more than a statement of the obvious. But imposing the requirement of invariance on our scientific theories proves to be an astonishingly fruitful thing to do. We have just described two different forms of invariance. The requirement that the laws of nature will not change if we spin around and determine them while facing different directions is called rotational invariance. The requirement that the laws will not change if we move from place to place is called translational invariance. These seemingly trivial requirements turned out to be astonishingly powerful in the hands of Emmy Noether, whom Albert Einstein described as the most important woman in the history of mathematics. In 1918 Noether published a theorem that revealed a deep connection between invariance and the conservation of particular physical quantities. We will have more to say about conservation laws in physics later on, but for now let us just state the deep result Noether discovered. For the specific example of looking at the world in different directions, if the laws of nature remain unchanged irrespective of the direction in which we are facing, then there exists a quantity that is conserved. In this case, the conserved quantity is called angular momentum. For the case of translational invariance, the quantity is called momentum. Why should this be important? Let's pull an interesting physics fact out of the metaphorical hat and explain it.

The moon moves 4 centimeters farther away from the earth every year. Why? Picture the moon in your mind's eye as being stationary above the surface of the spinning earth. The water in the oceans directly beneath the moon will bulge out a tiny bit toward the moon because the moon's gravity is pulling it, and the earth will rotate once a day beneath this bulge. This is the cause of the ocean tides. There is friction between the water and the surface of the earth, and this friction causes the earth's rate of spin to slow down. The effect is tiny but measurable; the earth's day is gradually lengthening by approximately two-thousandths of a second per century. Physicists measure the rate of spin using angular momentum, so we can say that the angular momentum of the earth is reducing over time. Noether tells us that because the world looks the same in every direction (to be more precise, the laws of nature are invariant under rotations), then angular momentum is conserved, which means that the total amount of spin must not change. So what happens to the angular momentum the earth loses by tidal friction? The answer is that it is transferred to the moon, which speeds up in its orbit around the earth to compensate for the slowing down of the earth's rotation. This causes it to drift slightly farther away from the earth. In other words, to ensure that the total angular momentum of the earth and moon system is conserved, the moon must drift into a wider orbit around the earth to compensate for the fact that the earth's rate of spin is slowing down. This is a very real and quite fantastic effect. The moon is big, and it is drifting farther away from the earth as every year goes by to conserve angular momentum. Italian novelist Italo Calvino found it so wonderful that he wrote a short story called "The Distance of the Moon," in which he imagined a time in the distant past when our ancestors could sail each night across the ocean in boats to meet the setting moon and clamber onto its surface using ladders. As the moon drifted farther away over the years, there came a night when the moon lovers had to make a choice between becoming trapped on the moon forever or returning to Earth. This surprising and, in the hands of Calvino, strangely romantic phenomenon has its explanation in the abstract concept of invariance and the deep connection between invariance and the conservation of physical quantities.

It is difficult to overstate the importance of the idea of invariance in modern science. At the heart of physics is the desire to produce an intellectual framework that is universal and in which the laws are never a matter of opinion. As physicists, we aim to uncover the invariant properties of the universe because, as Noether well knew, these lead us to real and tangible insights. Identifying the invariant properties is far from easy, however, because nature's underlying simplicity and beauty are often hidden.

Nowhere in science is this truer than in modern particle physics. Particle physics is the study of the subatomic world; the quest for the fundamental building blocks of the universe and the forces of nature that stick them together. We have already met one of the fundamental forces, electromagnetism. Understanding it led us to an explanation for the nature of light that has launched us on the road to relativity. In the subatomic world there are two other forces of nature that hold sway. The strong nuclear force sticks the atomic nucleus together at the heart of the atom, and the weak nuclear force allows stars to shine and is responsible for certain types of radioactive decay; the use of radiocarbon dating to measure the age of things, for example, relies on the weak nuclear force. The fourth force is gravity, the most familiar perhaps, but by far the weakest. Our best theory of gravity today is still Einstein's general theory of relativity and, as we shall see in the final chapter, it is a theory of space and time. These four forces act between just twelve fundamental particles to build everything in the world we can see, including the sun, moon, and stars, all the planets in our solar system, and indeed our own bodies. This all constitutes an astonishing simplification of what at first glance appears to be an almost infinitely complicated universe.

Glance out your window. You may be faced with the distorted reflections of a city, as the afternoon light scatters off sheets of steel and glass, or black and white cattle grazing in neatly fenced green fields. But whether cityscape or farmland, the most astonishing thing about practically every window view in the world is the evidence of human intervention. Our civilization is all-pervasive, and yet twenty-first-century physics tells us that, at its heart, it is all a mathematical dance involving a handful of subatomic particles, organized by only four forces of nature over 13.7 billion years. The complexity of human brains and the products of the powerful synthesis between consciousness and dexterous skill that we glimpse outside our windows mask the underlying simplicity and elegance of nature. The scientist's task is to hunt for those properties that act as a Rosetta stone, to allow us to decipher the language of nature and reveal its beauty.

The tool that allows us to search for and exploit these properties of nature is mathematics. In itself, this is a sentence that throws up deep questions, and entire books have been written attempting to advance plausible reasons as to why it may be so. Quoting Eugene Wigner again: "The miracle of the appropriateness of the language of mathematics for the formulation of the laws of physics is a wonderful gift which we neither understand nor deserve." Perhaps we will never understand the true nature of the relationship between mathematics and nature, but history has shown that mathematics allows us to organize our thinking in a way that proves to be a reliable guide to a deeper understanding.

As we have been at pains to emphasize, to proceed in this spirit, physicists write down equations, and equations do nothing more than express relationships between different real-world "things." An example of an equation is speed = distance/time, which we met in the last chapter when we were discussing light clocks: in symbols $v = x/t$, where v is the speed, x is the distance traveled, and t is the time taken to travel the distance x. Very simply, recall that if you travel 60 miles in 1 hour, then you have traveled at a speed of 60 miles per hour. Now, the most interesting equations will be those that are capable of furnishing a description of nature that is agreed upon by everyone. That is, they should deal *only* in invariant quantities. We could all then agree on what we are measuring, irrespective of our perspective in the universe. According to common sense, the distance between any two points in space is such an invariant quantity, and pre-Einstein it was. But we saw in the previous chapter that it is no such thing. Remember: Common sense is not always reliable. Similarly, the passage of time has become a subjective thing and it varies depending on how fast clocks are moving relative to each other. Einstein has upset the order of things, and we cannot even rely on distance and time to build a reliable picture of the universe. From the point of view of a physicist looking for the deep laws of nature, the equation $v = x/t$ is therefore of no fundamental use, because it does not express a relationship between invariant quantities. By undermining space and time, we have shaken the very foundations of physics. What, then, are we to do?

One option is to try and reestablish order by making a conjecture. Conjecture is a fancy word for "guess," and scientists do it all the time—there are no prizes for how smart we are in figuring out the underlying theory; a successful educated guess will do just fine so long as it agrees with experiment. The conjecture is radical: *Space and time can be merged into a single entity that we call "spacetime," and distances in spacetime are invariant.* This is a bold assertion and its content will become clearer as we go. When you think about it for a moment, it is perhaps less bold than it seems at first sight. If we are to lose the age-old certainties of absolute, unvarying distances in space and the unchanging tick-tock of time as the great clock in the sky marks the passing of the years, then maybe the only thing to do is to search for some kind of unification of the two seemingly separate concepts. Therefore, our immediate challenge is to search for a new measure of distance in spacetime that does *not* change depending on how we move around relative to each other. We will need to tread carefully to understand how the spacetime synthesis works. But what exactly does it mean to search for a distance in spacetime?

Suppose I get out of bed at 7 a.m. and finish my breakfast at 8 a.m. The following statements are true given what we know from experiment: (1) I may measure the distance in space from my bed to my kitchen to be 10 meters, but someone whizzing by at high speed will measure a different distance; (2) My watch indicates that I took 1 hour to eat breakfast, but the high-speed observer will record a different time. Our conjecture is that the distance in *spacetime* between my getting out of bed and my finishing breakfast is something we can all agree upon—i.e., it is invariant. The existence of this consensus is crucial because we want to build up a set of natural laws using only this type of object. Of course, we just guessed that this might be how things are and we certainly haven't proven anything yet. We haven't even decided how to calculate distances in spacetime. But to proceed further, we must first explain what is meant by the second of our three key words, *causality*.

Causality is another seemingly obvious concept whose application will have profound consequences. It is simply the requirement that cause and effect are so important that their order cannot be reversed. Your mother caused your birth, and no self-consistent picture of space and time should allow you to be born before your mother. To construct a theory of the universe in which you could be born first would be nonsense and lead to contradictions. When put in these terms, nobody could argue with the requirement of causality.

It is worth reflecting, however, that humans seem capable of ignoring it on a daily basis. Take prophesy, for example. Figures like Nostradamus are revered to this day for allegedly being able to see events that happen in the future, either in dreams or some other mystical trancelike state. In other words, events that happened centuries after Nostradamus' death were visible in his lifetime, at least to him. Nostradamus died in 1566, but he is credited with observing the Great Fire of London in 1666, the rise of Napoleon and Hitler, the September 11, 2001, attacks on the United States, and, our own personal favorite, the rise of the Antichrist in Russia in 1999. The Antichrist hasn't appeared yet but perhaps he/she is still rising and if he/she does appear before this book goes to print, then we stand corrected.

Putting amusing drivel aside, we need to introduce some important terminology. Nostradamus's death was an "event," as were the birth of Adolf Hitler and the Great Fire of London. For Nostradamus to observe an event such as the Great Fire that happened after his death would require the ordering of the two events to be reversed. To say this explicitly is almost a tautology; Nostradamus died before the Great Fire, and therefore he could not have observed it. To observe it, the event that is the Great Fire must have been available for viewing before the event that is Nostradamus's death, and therefore the order of the events must have been reversed. There is an important subtlety: Nostradamus could have caused the Great Fire. We could imagine that he left a sum of money in a bank account that encouraged someone to light a fire in Pudding Lane shortly after midnight on September 2, 1666. This would establish a causal link between the events associated with the life and death of Nostradamus and the events associated with the Great Fire of London. As we shall see later, it is in fact only the ordering of such connected events (called causally connected events) that cannot be reversed—cause and effect are sacred in Einstein's universe.

Other events occur far enough away from each other in space and time that they could not have any possible influence on each other. Remarkably, the ordering of these can be reversed. Einstein's theory exploits a loophole that allows the order of events to be switched provided that doing so makes absolutely no difference to the workings of the universe. We shall explain what we mean by "far enough away" later on. For now, we have introduced the concept of causality as an axiom that we shall use to build our theory of spacetime. The success of the theory in predicting the outcome of experiments will of course be the ultimate arbiter. As an aside, Nostradamus did get one prediction right. While suffering from a particularly acute bout of gout, he apparently told his secretary, "You will not find me alive at sunrise." The next morning he was found dead on the floor.

What has causality got to do with spacetime and, in particular, distances in spacetime? Well, we will soon discover that insisting on a causal universe constrains the structure of spacetime to such an extent that we are left with no choice in the matter. There will be only one way in which we can merge space and time together to manufacture spacetime while simultaneously preserving the causal order of things. Any other way would violate causality and allow us to do fantastical things like going

FIGURE 3

back in time to prevent our own birth or, in Nostradamus's case, perhaps avoiding a lifestyle that made him susceptible to gout.

Time now to return to the challenge of developing the concept of distance in spacetime. To get warmed up we will set time to one side for the moment and think about the idea of distance in ordinary three-dimensional space, a concept with which we are all familiar. Suppose we try to measure the shortest distance between two cities on a flat map of the earth. As will be very familiar to anyone who has flown on a long-haul flight and watched her progress on the map on the aircraft entertainment system, the shortest distance between any two points on the earth's surface appears as a curve. This line is known as a great circle. Figure 3 shows a map of the earth, and drawn on it is a line that corresponds to the shortest distance between Manchester and New York. On a globe, this line can be understood but at first glance it is a surprise to see a curved line representing the shortest distance between two points. This occurs because the earth's surface is not flat, but curved. To be specific, the earth is a sphere. The curved nature of the earth's surface is also the reason why, on some flat maps, Greenland looks much bigger than Australia, when in reality it is much smaller. The message is clear: Straight lines represent the shortest distance between two points only in flat space. The geometry of flat space is often called Euclidean geometry. What Euclid didn't know at the time, however, and in fact it did not become clear until the nineteenth century, was that his geometry of flat space is only a specific example of a whole family of different possible geometries, each of which are mathematically consistent and some of which can be used to describe nature. A very good example is the surface of the earth, which is curved and therefore described using a geometry that is non-Euclidean. Specifically, the shortest distance between two points is not a Euclidean straight line.

There are other familiar Euclidean properties that are not obeyed on the surface of the earth. For example, the interior angles of a triangle no longer add up to 180 degrees, and lines that are parallel and point north-south at the equator cross at the poles. If Euclid is no use anymore, we need to figure out how to calculate distances in a curved space, such as on the earth's surface. One way would be to work directly with a globe and measure out the distances using a piece of string. Now we would be correctly accounting for the curvature of the earth. An airline pilot could stretch a piece of string between two cities on the globe, measure its length with a ruler, and then simply multiply the answer by the ratio in size of the globe and the earth. But maybe we don't have a globe on hand, or maybe we need to write the computer software that helps airplanes navigate. In either case, we need to do better than a piece of string and figure out an equation that tells us the distance between any two points on the earth's surface given only their latitude and longitude, and the shape and size of the earth. Such an equation is not too hard to find and if you know a little mathematics you might even try to find it. We don't need to write it down here, but the point is that an equation exists and it hasn't got much to do with the Euclidean geometry of a flat tabletop. It does, however, allow one to calculate the shortest distance between two points on a sphere, in much the same way that Pythagoras' theorem is a recipe for calculating the shortest distance between two points (the hypotenuse) on a tabletop if we know the distances from one corner as measured along the edges of the table. Since straight lines belong in the domain of Euclid, we shall introduce a new term for the shortest distance between two points that applies whether the space is curved or flat. This line is called a geodesic: A great circle is a geodesic on the surface of the earth and a straight line is a geodesic in flat space. So much for distances in three-dimensional space. Now we must decide how to measure distances in spacetime, so let's go ahead and complicate matters by adding time into the mix.

We already introduced the concepts we will need when we thought about getting out of bed and finishing breakfast in the kitchen. There is no problem in saying that the distance in space between the bed and the kitchen is 10 meters. We could also say, although it sounds rather strange, that the distance in time between getting out of bed and finishing breakfast is 1 hour. This is not how we naturally think about time, because we are not used to describing it in the language of geometry. We would rather say "one hour passed between my getting out of bed and finishing my breakfast." In the same way, we would not normally say "10 meters have passed since I got out of bed and sat down in the kitchen." Space is space, and time is time, and never the twain shall be intermingled. But we have set ourselves the task of trying to merge space and time together, because we suspect that this is the only way to rebuild things in a way that fits with Maxwell and Einstein. So let us proceed and see where it leads us. If you are not a scientist, then this may be the most difficult part of the book so far because we are operating in a purely abstract fashion. The capacity for abstract thought is what gives science its power, but also perhaps gives it a reputation as being difficult because it is not a faculty we generally need too much in everyday life. We have already encountered a difficult abstract concept in the form of the electric and magnetic fields, and in fact the abstraction needed to merge space and time together is probably less challenging than that.

What we are doing implicitly in speaking of "the distance in time" is treating time as an additional dimension. We are used to the phrase "3-D," as in three-dimensional, referring to the fact that space has three dimensions: up and down; left and right; forward and backward. When we try to add time into the framework, so that we can define distances in spacetime, we are in effect creating a four-dimensional space. To be sure, the time dimension behaves differently than the space dimensions. We have complete freedom of movement in space, whereas we go only one way in time, and time doesn't feel anything like space. But that need not be an insurmountable hurdle. Thinking of time as "just another dimension" is the abstract leap we have to take. The trick, if it sounds too confusing, is to imagine how you might feel if you were a creature that could only ever move forward, backward, left and right. You have never experienced up and down—you live in a flat world. If someone asked you to imagine a third dimension, your flat mind would not be able to grasp it. But if you had a mathematical bent, you might be happy to accept the possibility and, in any case, you could still do the maths even if you couldn't picture the mysterious extra dimension in your mind's eye. Likewise for human beings and four-dimensional space. It should become more natural to think of time as "just another dimension" as our story unfolds. If there is one thing we try to teach our students when they first arrive at the University of Manchester, ready to learn to be physicists, it is that everyone gets confused and stuck. Very few people understand difficult concepts the first time they encounter them, and the way to a deeper understanding is to move forward with small steps. In the words of Douglas Adams: "Don't panic!"

Let us continue in a gentler vain for a moment by noticing something very simple: Things happen. We wake up, we make breakfast, we eat breakfast, and so on. We'll call the occurrence of a thing "an event in spacetime." We can uniquely describe an event in spacetime by four numbers: three spatial coordinates describing where it happened and a time coordinate describing when it happened. Spatial coordinates can be specified using any old measuring system. For example, longitude, latitude, and altitude will do if the event is occurring in the vicinity of the earth. So your coordinates in bed might be N 53° 28' 2.28", W 2° 13' 50.52", and 38 meters above sea level. Your time coordinates are specified using a clock (because time is not universal, we'll have to say whose clock in order to be unambiguous) and might be 7 a.m. GMT when your alarm goes off and you wake up. So we have four numbers that uniquely locate any event in spacetime. Notice that there is nothing special about the particular choice of coordinates. In fact, these particular coordinates are measured relative to a line passing through Greenwich in London, England. This convention was agreed upon in October 1884 by twenty-five nations, with the only dissenting voice being San Domingo (France abstained). It is a very important concept that the choice of coordinates should make absolutely no difference.

Let's take the moment when I wake up in bed as our first event in spacetime. The second event could be the event that marks the end of breakfast. We have said that the spatial distance between the two events is 10 meters and the distance in time is 1 hour. To be unambiguous we'd need to say something like "I measured the distance between my bed and my breakfast table using a tape measure whose ends were stretched directly from bed to table" and "I measured the time interval using my bedside clock and the clock sitting in my kitchen." Don't forget that we already know that these two distances, in space and in time, are not universally agreed upon. Someone flying past your house in an aircraft would say that your clock runs slow and the distance between your bed and your breakfast table shrinks. Our aim is to find a distance in spacetime upon which everyone agrees. The million-dollar question is then "how do we take the 10 meters and the 1 hour to construct an invariant distance in spacetime?" We need to tread carefully and, just like in the case of distances on the earth's surface, we shall not assume Euclidean geometry.

If we are to compute distances in spacetime, then we have an immediate problem to resolve. If distance in space is measured in meters and distance in time in seconds, how can we even begin to contemplate combining the two? It is like adding apples and oranges, because they are not the same type of quantity. We can, however, convert distances into times and vice versa if we use the equation we met earlier, $v = x/t$. With a miniscule bit of algebra we can write time $t = x/v$, or distance $x = vt$. In other words, distance and time can be interchanged using something that has the currency of a speed. Let us therefore introduce a calibrating speed; call it c. We can then measure time in meters provided we take any time interval and multiply it by our calibrating speed. At this point in our reasoning c really can be any old speed and we have not committed ourselves at all as to its actual value. Actually, this trick of interchanging time and distance is very common in astronomy, where the distance to stars

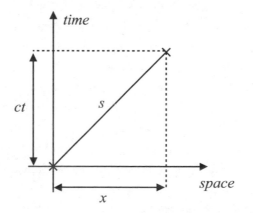

FIGURE 4

and galaxies is often measured in light-years, which is the distance light travels in one year. This doesn't seem so strange because we are used to it, but it really is a distance measured in years, which is a unit of time. In the astronomy case, the calibrating speed is the speed of light.

This is progress; we now have time and distance intervals in the same currency. For example, they could both be given in meters, or miles or light-years or whatever. Figure 4 illustrates two events in spacetime, denoted by little crosses. The bottom line is that we want a rule for figuring out how far apart the two events are in spacetime. Looking at the figure, we want to know the length of the hypotenuse given the lengths of the other two sides. To be a little more precise, we shall label the length of the base of the triangle as x while the height is ct. It means that the two events are a distance x apart in space and a distance ct apart in time. Our goal, then, is to answer the question "what is the hypotenuse, s, in terms of x and ct?" Making contact with our earlier example $x = 10$ meters is the distance in space from bed to kitchen table, and $t = 1$ hour is the distance in time. So far, since c was arbitrary, ct can be anything and we appear to be treading water. We shall press onward nonetheless.

We have to decide on a means of measuring the length of the hypotenuse, the distance between two events in spacetime. Should we choose Euclidean space, in which case we can use Pythagoras' theorem, or something more complicated? Perhaps our space should be curved like the surface of the earth, or maybe some other more complicated shape. There are in fact an infinite number of ways that we might imagine calculating distances. We'll proceed in the way that physicists often do and we will make a guess. Our guess will be guided by a very important and useful principle called Occam's razor, named after the English thinker William of Occam, who lived at the turn of the fourteenth century. The idea is simple to state but surprisingly difficult to implement in everyday life. It might be summarized as "don't overcomplicate things." Occam stated it as "plurality must never be posited without necessity," which does beg the question: Why didn't he pay more attention to his own rule when constructing sentences? However it is stated, Occam's razor is very powerful, even brutal, when applied to reasoning about the natural world. It really says that the simplest hypothesis should be tried first, and only if this fails should we add complication bit by bit until the hypothesis fits the experimental evidence. In our case, the simplest way to construct a distance is to assume that at least the space part of our spacetime should be Euclidean; in other words, space is flat. This means that the familiar way of working out the distance in space between objects in the room in which we are seated reading this book is carried over into our new framework intact. What could be simpler? The question, then, is how we should add time. Another simplifying assumption is that our spacetime is unchanging and the same everywhere. These are important assumptions. In fact, Einstein did eventually relax them and doing so allowed him to contemplate the mind- (and space-) bending possibility that spacetime could be constantly changed by the presence of matter and energy. It led to his general theory of relativity, which is to this day our best theory of gravity. We will meet general relativity in the final chapter, but for the moment we can ignore all these twists and turns. Once we follow Occam and make these two simplifying assumptions, we are left with only two possible choices as to how to calculate distances in spacetime. The length of the hypotenuse *must* be either $s^2 = (ct)^2 + x^2$ or $s^2 = (ct)^2 - x^2$. There is no other option. Although we did not prove it, our assumption that spacetime should be unchanging and the same everywhere leads to only these two possibilities and we must pick either the plus sign or the minus sign. Of course, proof or no proof, we can be pragmatic and see what happens when we try each one on for size.

Flipping the sign means that the mathematics is not much of an extension over the by now familiar equation of Pythagoras. Our task is to figure out whether we should stick with the plus-sign version of Pythagoras, or shift to the minus-sign version of the distance equation. This may look at first sight to be a rather odd thing to investigate. What possible reason could there be for even considering Pythagoras with a minus sign? But that is not the right way to think about things. The formula for distances on a sphere looks nothing like Pythagoras either, so all we are doing is entertaining the idea that spacetime might not be flat in the sense of Euclid. Indeed, since the minus-sign version is the only option other than the plus-sign version (given our assumptions), we have no logical reason to throw it out at this stage. We should therefore keep it and investigate the consequences. If neither the plus- nor the minus-sign versions do the job, and we fail in constructing a workable distance measure in spacetime, then we must go back to the drawing board.

We are now about to plunge into a very elegant but perhaps tricky piece of reasoning. We will stick to our promise of using nothing more complicated than Pythagoras, but you might find that you have to read it twice. It should be worth it, because if you follow closely you might experience a feeling described by biologist Edward O. Wilson as the Ionian Enchantment. It derives from the work of Thales of Miletus, who is credited by Aristotle, two centuries later, as laying the foundations of the physical sciences in Ionia in the sixth century BCE. This poetic term describes the belief that the complexity of the world can be explained by a small number of simple natural laws because at its heart it is orderly and simple (we are reminded of Wigner's essay). The scientist's job is to strip away the complexity we see around us and to uncover this underlying simplicity. When the process works out, and the simplicity and unity of the world are revealed, we experience the Ionian Enchantment. Imagine for a moment cradling a snowflake in the palm of your hand. It is an elegant and beautiful structure, possessed of a jagged crystalline symmetry. No two snowflakes are alike, and at first sight this chaotic state of affairs seems to defy a simple explanation. Science has taught us that the apparent complexity of snowflakes hides an exquisite underlying simplicity; each is a configuration of billions of molecules of water, H_2O. There is nothing more to a snowflake than that, and yet an overwhelming complex of structure and form emerges when those H_2O molecules get together in the atmosphere of our planet on a cold winter's night.

To settle the question of the plus or minus sign, we need to turn our attention to causality. Let us first suppose that Pythagoras' is the right equation for distances in spacetime—i.e., $s^2 = (ct)^2 + x^2$. Yet again we return to our two events: waking up in bed at 7 a.m. and finishing breakfast in the kitchen at 8 a.m. We'll do something that may send shivers up your spine as you remember sitting in mathematics classes at school and gazing out the window across the football fields, pristine and inviting in the spring afternoon sunlight—let the waking-up event be called O and the finishing-breakfast event be called A. We do this purely for reasons of brevity, without wishing to don tweed and cover ourselves in chalk dust.

We know that the spatial distance between O and A is $x = 10$ meters and the distance in time between the two events is $t = 1$ hour, where x and t are measured by me. We haven't decided what c is yet, but when we do we will know ct and we can then go ahead and use the distance equation to calculate s, the distance in spacetime between events O and A. Our hypothesis is that while x and t can and will be different if they are measured by someone flying past at close to the speed of light, the distance s will stay the same. In other words, x and t can and will change but they must change in such a way that s never changes. To risk overemphasizing the point, we want to remind you that our goal is always to build the laws of physics using invariant objects in spacetime and the distance s is just such an object. If that sounds too abstract, then we can say it again but this time using less mathematically fancy language: Nature's rules must express relationships between real things, and those things live in spacetime. A thing living in spacetime is akin to an object sitting in a room. Spacetime (or the room) is the arena in which the thing lives. The nature of real things is not a matter of opinion and in that sense we say they are invariant. A three-dimensional example of something that is not an invariant might be the flickering shadow of an object sitting in a room illuminated by a warming fire. Clearly the shadow varies depending on how

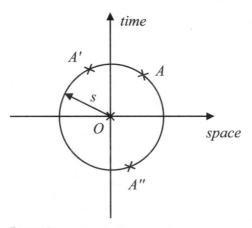

FIGURE 5

the fire is burning and where the fire is but we are never in any doubt that a real, unvarying object is responsible for it. Using spacetime, our plan is to lift physics out of the shadows and hunt down relationships between real objects.

The fact that two different observers can disagree on the values of x and t, provided s is the same, has a very important consequence, which can be visualized quite simply. Figure 5 shows a circle centered on O, the waking-up event, with a radius s. Because we are, for the moment, using the Pythagorean form of the distance equation, every point on the circumference of the circle is the same distance s away from O. This is a pretty obvious statement: The distance s is the radius of the circle. Points outside the circle are farther away from O while points inside are closer to O. But our hypothesis is that s is the distance in spacetime between events O and A. In other words, the event A could lie anywhere on the circumference of the circle and still

be a distance *s* in spacetime from *O*. At what point on the circle
should event *A* lie? That depends on who is measuring *x* and *t*.
For me in the house, we know exactly where it should be since
$x = 10$ meters and $t = 1$ hour. This is what we have drawn on
the diagram and labeled *A*. For a person flying past in a high-
speed rocket, the distance *x* in space and the distance *t* in time
will change, but if *s* is to remain the same, then the event must
still lie somewhere on the circle. So different observers record
different positions in space and time separately for the same
event, but subject to the constraint that we only slide the point
around on the circle. We've labeled two possible positions *A'* and
A''. For position *A'*, nothing particularly interesting has happened, but look carefully at position *A''*. Something very dramatic indeed has happened. *A''* has a negative distance in time
from *O*. In other words, *A''* happened before *O*. It is now in the
O's past. This is a world where you finish your breakfast before
you wake up! Such a circumstance is a clear violation of our
cherished axiom of causality.

As an aside, pictures like the ones shown in Figures 4 and 5
are called "spacetime diagrams" and they often help us work out
what is going on. They really are simple things. Crosses on a
spacetime diagram denote events and we can drop a line down
onto the line marked "space" (the space axis) from the event to
work out how far apart in space the event lies from the event *O*.
Likewise, a horizontal line drawn to the line marked "time" (the
time axis) tells us the time difference between the event and the
event *O*. We can interpret the area above the space axis as the
future of *O* (because *t* is positive for any event in this region)
and the area below as the past (because *t* is then negative). The

problem we have encountered is that we have constructed a definition of the distance in spacetime s between the events O and A that allows for A to be in either the future or the past of O, depending on how the person who observes the events is moving. In other words, we have discovered that the requirement of causality is intimately related to the way that we define the distance in spacetime, and the simple Pythagorean definition with the plus sign is no good.

We are faced with what the English biologist Thomas Henry Huxley famously described as "the great tragedy of science— the slaying of a beautiful hypothesis by an ugly fact." Huxley, known as Darwin's bulldog for his sterling defense of evolution, was once asked by William Wilberforce whether it was from his grandfather or grandmother that he claimed his descent from a monkey. Huxley is said to have replied that he would not be ashamed to have a monkey for his ancestor, but he would be ashamed to be connected with a man who used his great gifts to obscure the truth. The tragic truth in our case is that we must reject the simplest hypothesis if we are to preserve causality, and move on to something a little more complicated.

Our next and in fact only remaining hypothesis is that the distance between points in spacetime is to be calculated using $s^2 = (ct)^2 - x^2$. In contrast to the plus-sign version, this is a world where Euclidean geometry does not apply, as in the case of geometry on the surface of the earth. Mathematicians have a name for a space in which the distance between two points is governed by this equation: It is called hyperbolic space. Physicists have a different name for it. They call it Minkowski spacetime. The reader might take this to be a clue that we are on the

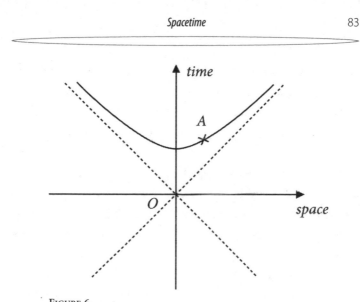

FIGURE 6

right track! Our top priority must be to establish whether Minkowski spacetime violates the demands of causality.

To answer this question we need once again to take a look at the lines in spacetime that lie a constant distance *s* from *O*. That is, we want to consider the analogue of the circles in Euclidean spacetime. The minus sign makes all the difference. Shown in Figure 6 are the same old events, *O* and *A*, along with the line of points that lie the same spacetime distance *s* from *O*. Crucially, these points no longer lie on a circle. Instead they lie on a curve known to mathematicians as a hyperbola. Mathematically speaking, all the points on the curve satisfy our distance equation—i.e., $s^2 = (ct)^2 - x^2$. Notice that the curve tends toward the dotted straight lines that lie at 45 degrees to the axes. Now the situation as viewed by observers in rocket ships is completely different from the plus-sign version because event *A* always stays in the future of event O. We can slide A around but never into O's past. In other words, everyone agrees that we wake up before we finish our breakfast. We can breathe a sigh of relief: Causality is not violated in Minkowski spacetime.

It's worth repeating this because it is one of the most important points in the book. If we decide to define the distance in spacetime between the two events O and A using Pythagoras' equation but with a minus sign, then no matter how anyone views the two events, A never crosses into O's past; it just moves around on the hyperbola. This means that if event A is in O's future according to one observer, then every other observer will also agree that A is in O's future too. Because the hyperbola never ever crosses into O's past, everyone agrees that eating breakfast comes after waking up.

We've just completed a subtle piece of reasoning. It certainly does not mean that we are correct in our original hypothesis that there should be an "invariant" distance in spacetime that is agreed upon by all observers. What it does mean, though, is that our hypothesis has survived an important test—it has survived the demands of the requirement of causality. We are not finished, however, because we are not just playing around with mathematics. We are physicists, and we are trying to construct a theory that describes how the world works. The ultimate and decisive test of our theory will be whether it can produce predictions that agree with experiment, and we are not yet ready to make a prediction, because we don't know what the calibrating speed c is. Without a number, we simply can't do the sums.

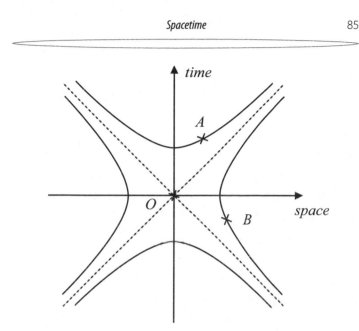

Remember, we needed *c* in order to have any chance of defining the notion of distance in spacetime, because we had to measure space and time in the same currency, but so far we have no idea what it actually represents. Is it the speed of anything interesting? The key to the answer lies in an intriguing property of the Minkowski spacetime we have just constructed. Those lines at 45 degrees are important. In Figure 7 we've drawn several other curves, each of constant spacetime distance from *O*. The important point is that there are in fact four types of curve that we can draw. One lies wholly in the future of event *O*, one lies always in the past, and two others lie to the left and right. They look a little bit worrying because they cross the horizontal line in just the same way that our circles crossed it in the case of the plus-sign version of Pythagoras. In the plus-sign case, this led us

to reject the hypothesis because it meant that causality was violated. Are we in the same boat with the minus-sign version? Are
we sunk? Well, no, there is a way out. Figure 7 shows an event B
sitting in the troubling region. It lies in O's past according to the
figure. But the hyperbola of constant distance from O for this
event crosses the space axis, with the implication that it is possible for some observers to consider event B as occurring in O's
future, while for others it is in O's past. Don't forget: Every observer must agree on the spacetime distance between events
even if they do not agree on the distances in space and time separately. It looks like a breakdown of causality, but fortunately
that is very definitely not the case.

How are we to restore causality to our theory of spacetime?
To answer this question, we need to think a little more carefully
about what we mean by causality. This next piece will involve
rocket ships and lasers, so if the abstract reasoning of the previous sections has left you drained, then you can relax for a
while. Let's think about event O again: waking up in bed in the
morning. To be a little more precise, the event could correspond
to my alarm clock going off. Shortly beforehand, on a planet in
the Alpha Centauri system, the nearest star system to Earth at a
distance of just over 4 light-years, a spaceship lifts off and heads
toward Earth. Must everyone agree that the spaceship started
its journey before I woke up? From the point of view of causality the issue depends critically upon whether information can
travel infinitely fast or not. If information can travel infinitely
fast, then the alien spaceship might conceivably be able to fire a
laser beam that travels in an instant to the earth and destroys my
alarm clock. The result is that I oversleep and miss breakfast.

Missing breakfast might be the least worrying issue given this particular scenario, but we are doing a thought experiment, so let us ignore the emotional consequences of having our alarm clock vaporized by an alien laser and continue. The firing of the spaceship's laser caused me to miss breakfast, and therefore the ordering cannot be swapped without violating our doctrine of the protection of causality. This is easy to see because if some observer were able to conclude that the spaceship took off after I woke up, then we would have a contradiction because I cannot oversleep if I have already woken up. We are forced to conclude that if information can travel at arbitrarily high speeds, then it can *never* be permissible to switch the time ordering of any two events without violating the law of cause and effect. But there is a loophole in our reasoning that permits the time ordering of certain pairs of events to be flipped, but only if they lie outside the 45-degree lines. These lines are beginning to look very important indeed.

Let us imagine the alien-laser-exploding-alarm-clock incident again, but now subject it to a cosmic speed limit. That is to say, we will not allow the laser beam to travel infinitely fast from the spaceship to our alarm clock. Covering ourselves in a thin mist of chalk dust for the last time, we call the laser-firing event B, as illustrated in Figure 7. If the spaceship fired the laser (event B) very shortly before the alarm clock–ringing event O, from a very great distance away, then there is no way the spaceship could possibly prevent me from waking up because the laser beam simply hasn't got enough time to travel from the ship to my clock. This must be the case if the laser beam is constrained to travel at or below some kind of cosmic speed limit. If this is the situation, the events O and B are said to be causally disconnected.

As illustrated in the figure, we are supposing that B happens just before O such that it lies in the right-hand wedge region, which is the "dangerous" region for causality. Different observers will generally disagree on whether B happens before or after O, because their different points of view correspond to moving B around on the hyperbola, which crosses the space axis from the future to the past. This is unavoidable, but cause and effect can still be protected if there is absolutely no way that event B can influence event O. In other words, who cares whether B happened in O's past or future, if it makes no difference to anything because B and O cannot influence each other? There are four distinct regions in Minkowski spacetime, separated from each other by the 45-degree lines. If we are to protect causality, then any event that occurs in either of the left-hand or right-hand wedges must never be able to send a signal that can possibly reach O.

To interpret the delineating lines, look again at our spacetime diagrams. The horizontal axis represents distance in space, and the vertical axis represents distance in time. The 45-degree lines therefore correspond to events that have a distance in space from O that is equal to the distance in time (ct). How fast must a signal travel from O if it is to influence an event lying exactly on the 45-degree line? Well, if the event is 1 second in O's future, then the signal must travel a distance c x 1 second. If it's 2 seconds in the future, then it must travel a distance c x 2 seconds. In other words, it must travel at the speed c. For a signal to travel between B and O, therefore, it must travel faster than the speed c. Conversely, for any events that lie between the 45-degree lines but in the upper and lower wedges, it is possible to communicate between them and the event at O using signals that travel at speeds slower than c.

We have finally managed to interpret the speed c: It is the cosmic speed limit. Nothing can travel faster than c because if it did it could be used to transmit information that could violate the principle of cause and effect. Notice also that if everyone is to agree on the distance in spacetime between any two events, then they must also agree that the cosmic speed limit is c, regardless of how they are moving around in spacetime. The speed c therefore has an additional interesting property: No matter how two different observers are moving, they must always measure c to be the same. The speed c is beginning to look a lot like another special speed we have encountered in this book: the speed of light, but we haven't proved the connection yet.

Our original conjecture is still very much alive. We have managed to build a theory of space and time that looks capable of reproducing the physics we met in the last chapter. Certainly, the existence of a universal speed limit offers promise, especially if we can interpret it as the speed of light. We also have a spacetime in which space and time are no longer absolutes. They have been sacrificed in favor of absolute spacetime. To convince ourselves that we have constructed a possible description of the world, let's see if we can obtain the slowing down of moving clocks that we met in Chapter 3.

Imagine that you are back on the proverbial train, sitting down in a carriage wearing a wristwatch. For you, it is convenient to measure distances relative to your own position and times using your wristwatch. Your train journey takes two hours from station to station. Since you never leave your seat throughout the journey, you have traveled a distance $x = 0$. This is the principle we established right at the start of the book. It is not possible to define who is moving and who is standing still, and therefore it is perfectly acceptable for you, seated on a train, to decide that you are not moving. In this case, only time passes. Since your journey takes two hours, then, from your perspective, you have traveled only in time. In spacetime, therefore, you have traveled distance s given by $s = ct$ where $t = 2$ hours (because the distance in space as measured by you is $x = 0$). That is all straightforward. Now consider your journey from the standpoint of your friend, who is not on the train but who instead is sitting on the ground somewhere (it does not matter where he actually is, just that he is at rest relative to the earth while you are whizzing by on the train). Your friend would prefer to measure times using his own wristwatch and distances relative to himself. To simplify things a little bit, let us suppose your train journey is on a perfectly straight track. If you travel for 2 hours at a speed of $v = 100$ miles per hour, then your friend notes that, at the end of the journey, you have traveled a distance $X = vT$. We are using capital letters when we talk about distances or times measured by your friend in order to distinguish them from the corresponding quantities measured by you (i.e., $x = 0$ and $t = 2$ hours). So, according to your friend, you have traveled a spacetime distance s given by $s^2 = (cT)^2 - (vT)^2$.

Here is the crucial part of the whole argument: You must both agree on the spacetime distance of your journey. According to your measurements, you did not move ($x = 0$) and your journey took 2 hours ($t = 2$ hours), while your friend says that you have traveled a distance of vT (where $v = 100$ miles per hour) and your journey takes a time T. Well, we are obliged to equate the corresponding distances in spacetime and so $(ct)^2 = (cT)^2 - (vT)^2$. This formula can be jiggled around to give us $T = ct/\sqrt{c^2 - v^2}$. So, although your wristwatch registers that your journey lasted for 2 hours, according to your friend your journey lasted a little longer. The enhancement factor is equal to $c/\sqrt{c^2 - v^2} = 1/\sqrt{1 - v^2/c^2}$, which is exactly what we got in the last chapter but only if we interpret c as the speed of light.

Are you beginning to feel the Ionian Enchantment? We have deduced the same formula that emerged from thinking about light clocks and triangles in the previous chapter. Then, we were motivated to think about light clocks because Maxwell's brilliant synthesis of the experimental results of Faraday and others strongly suggested that the speed of light should be the same for all observers. This conclusion was supported by the experimental work of Michelson and Morley, and taken at face value by Einstein. In this chapter we arrived at exactly the same conclusion but with no reference to history or experiment. We didn't even need to give light a special role. Instead, we introduced spacetime and, as a result, insisted that there should exist the notion of an invariant distance between events. On top of that we demanded that cause and effect be respected. We then constructed the simplest possible distance measure and remarkably arrived at the same answer as Einstein. This reasoning is perhaps one of the most beautiful examples of the unreasonable effectiveness of mathematics in the physical sciences. Thales would be so enchanted that he would already be reclining in a bath of asses' milk having been scrubbed by eunuchs. For his concubines to enter his bathroom carrying wine and figs, all we have to do is establish that c must be the speed of light using an argument that is entirely independent of the historical reasoning we encountered in the last chapter. That climax will arrive in the next chapter, for now we can take a rest from the maths, leave Thales poised in anticipation, and revel in the fact that we have succeeded in uncovering a whole new way of thinking about Einstein's theory. Spacetime really does seem to work—the notion of a unified space and time makes sense, just as Minkowski said.

How are we to picture spacetime? Real spacetime is four-dimensional but the four-dimensional nature poses a stumbling block to our imagination, because human brains cannot directly picture objects in higher than three dimensions. In addition, the fact that time makes up one of the dimensions just sounds plain weird. A picture that might help make it all a little less mystical is to imagine a motorcycle roaming over an undulating countryside. Roads criss-cross the landscape, allowing our motorcyclist to wander this way and that. Spacetime is rather like the rolling countryside. The analogue of our motorcyclist traveling due north might be an object moving only in the time direction through spacetime. In other words, the object would be stationary in space. Of course, statements like "stationary in space" are subjective and so it is to be understood that the identification of "due north" with "the time direction" implies a particular point of view, but that is okay; we just need to bear it in mind.

Now, the roads criss-crossing the spacetime landscape are all restricted to lie within a bearing of 45 degrees of north; roads due east and west are disallowed because to travel along them our spacetime "motorcyclist" would have to exceed the cosmic speed limit through space. Think of it this way: If the motorcyclist could travel due east, then he could go as far as he wanted in the easterly direction without any time passing at all, because he would not travel any distance up the northerly time direction. This would correspond to an infinite speed through space; he would get from a to b instantaneously. The roads have therefore been built so that the motorcyclist cannot travel too fast in an easterly or westerly direction.

The analogy can be pushed even further. We will very soon show that everything moves over spacetime at the same speed. It is just as if our motorcyclist has a device that fixes the throttle on his bike so that he always travels at the same speed over the spacetime landscape. We do need to be a little bit careful here, for when we talk about a speed in spacetime, it is not the same as a speed through space. A speed through space can be anything provided it does not exceed the cosmic speed limit— e.g., our motorcyclist might take a road close to a bearing of northeast, and in doing so he would be pushing as close to the cosmic speed limit as he could. In contrast, a road bearing close to due north would not lead to much movement east or west and consequently a journey that is well within the speed limit. The statement that everything moves at the same speed through spacetime sounds rather profound and perhaps a little baffling. It means that as you sit reading this book you are whizzing over the spacetime landscape at exactly the same speed as everything else in the universe. Viewed like that, motion through space is a shadow of a more universal motion through spacetime. In a very real sense, as we will now show, you are exactly like the motorcyclist with the fixed throttle. You are moving over the spacetime landscape with your throttle fixed open as you read this book. Because you are sitting still, your journey is entirely up the northerly time road. If you glance at your watch, you'll see the distance in time ticking by. This is a very strange-sounding claim, so let's go through it carefully.

Why does everything move at the same speed through spacetime? Consider our motorcyclist again and imagine 1 second passes according to the watch on his wrist. In that time, he will have traveled through spacetime by a certain distance. But everyone must agree on how far that distance is, because distances in spacetime are universal and not a matter for debate. That means we can ask the motorcyclist how far he thinks he has traveled over the spacetime landscape and the answer he gives will be the right answer. Now, the motorcyclist can choose to calculate distances in spacetime relative to himself, and from this point of view he has not moved in space. It is just like the person sitting on the airplane in Chapter 1 who doesn't stray from her airplane seat and who therefore states that she has not moved. She may have moved relative to someone else—for example, someone standing on the ground watching the plane fly by—but that is not the point. So from our motorcyclist's point of view, he has not moved in space and yet 1 second in time has passed. He can therefore use the spacetime distance equation $s^2 = (ct)^2 - x^2$ with $x = 0$ (because he hasn't moved in space) and $t = 1$ second to figure out how far in spacetime he has actually traveled: The answer is a distance equal to c multiplied by 1 second. So the motorcyclist tells us that he is traveling a distance of c (multiplied by 1 second) for every second that passes on his watch, and that is just another way of saying that his speed through spacetime is equal to c. If you have been following closely, then you might object that the passage of 1 second was measured on the motorcyclist's wristwatch and that a different amount of time will pass according to someone else who is moving relative to the motorcyclist. That is true enough, but there is something special about the motorcyclist's watch, because the motorcyclist does not move relative to himself (a trivial statement). We are therefore free to put $x = 0$ in the distance equation and so the time that passes on his wristwatch is a direct way to measure the spacetime distance s. This is a nice result: The time that passes on the motorcyclist's watch is equal to the spacetime distance traveled divided by c. In a sense, his watch is a device for measuring distances in spacetime. Since both the spacetime distance and c are agreed upon by everyone, it follows that the motorcyclist has unwittingly used his watch to measure something that everyone can agree upon. The spacetime speed c that he deduces is therefore also a quantity that everyone can agree upon.

So the speed through spacetime is a universal upon which everyone agrees. This newfound way of thinking about how things move through spacetime can help us get a different handle on why moving clocks run slow. In this spacetime way of thinking, a moving clock uses up some of its fixed quota of spacetime speed because of its motion through space and that leaves less for its motion through time. In other words, a moving clock doesn't move so fast through time as a stationary one, which is just another way of saying that it ticks more slowly. In contrast, a clock sitting at rest whizzes along in the time direction at the speed c with no motion through space. It therefore ticks along as fast as is possible.

Armed with spacetime, we are ready to contemplate one of the wonderful puzzles of Special Relativity: the Twins Paradox. Earlier in the book we showed that Einstein's theory allows us to contemplate the possibility of traveling to distant places in the universe. Speeding within a whisker of the speed of light, we imagined journeying off to the Andromeda galaxy within a human lifetime regardless of the fact that it takes light nearly 3 million years to make the journey. There is a paradox lurking here that we previously glossed over. Imagine twins, one of whom trains to be an astronaut and heads off on humanity's first mission to Andromeda, leaving her twin back home on Earth. The astronaut twin is moving at high speed relative to the earth and consequently her life slows down relative to her twin on Earth. But we have just spent a significant fraction of this book arguing that there is no such thing as absolute motion. In other words, the answer to the question "Who is doing the moving?" is "Whoever you want." Anybody and everybody is free to decide that they are standing still, and the other guy is whizzing around the universe at high speed relative to them. And so it is for the astronaut twin, who is free to say that she is standing perfectly still in her space rocket, watching the earth fly away at high speed. For her, it is therefore the earthbound twin who ages more slowly. Who is right? Can it really be that each of the twins ages more slowly relative to the other? Well it has

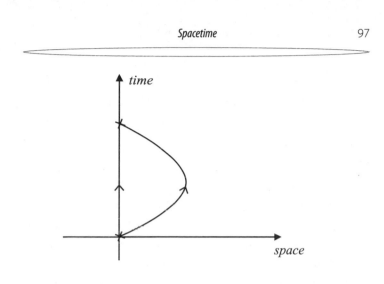

FIGURE 8

to be like that—that is what the theory says. There is no paradox yet, because any problems you might be having in believing that each twin observes the other to be aging more slowly are not real problems. They are due to the fact that you are clinging to the idea of universal time. But time is not universal; that much we have learned, and that means there is no contradiction at all. Now comes the apparent paradox: What happens if the astronaut twin returns back to Earth sometime in the future and meets up with her earthbound twin? Obviously they cannot both be younger than the other. What is going on? Is one of them actually older than the other? If so, who?

The answer can be found in our understanding of spacetime. In Figure 8 we show the paths through spacetime taken by the twins, as measured using clocks and rulers at rest relative to the earth. The earthbound twin stays on the earth and consequently her path snakes along the time axis. In other words, almost all of her allocated speed through spacetime is expended traveling

through time. Her astronaut twin, on the other hand, heads off
at close to light speed. Returning to the motorcyclist analogy,
that means she charges off in a "northeasterly" direction, using
up as much of her spacetime speed as she can to push through
space at close to the cosmic speed limit. On the spacetime diagram shown in Figure 8, that means she travels close to 45 degrees. At some point, however, she needs to turn around and
come back to the earth. The picture shows that we are supposing that she heads back again at close to light speed but this time
in a "northwesterly" direction. Obviously the twins take different paths through spacetime, even though they started and finished at the same point.

Now just like distances in space, the length of two different
paths in spacetime can be different. To reiterate, although everyone must agree on the length of any particular path through
spacetime, the lengths of different paths need not be the same.
This is really no different from saying that the distance from
Chamonix to Courmayeur depends upon whether you went
through the Mont Blanc tunnel or hiked over the Alps. Of
course, walking over a mountain means you travel a longer distance than tunneling through it. In our discussion of the motorcyclist speeding over the spacetime landscape, we established
that the time measured on the motorcyclist's wristwatch provides a direct way to measure the spacetime distance he traveled: we just need to multiply the elapsed time by c to get the
spacetime distance. We can turn this statement on its head and
say that once we know the spacetime distance traveled by each
of the twins, then we can figure out the time that passes according to each. That is, we can think of each twin as a voyager

through spacetime with their wristwatches measuring the spacetime distance that they travel.

Now comes the key idea. Look again at the formula for distances in spacetime, $s^2 = (ct)^2 - x^2$. The spacetime distance is biggest if we can follow a path that has $x = 0$. Any other path must be shorter because we have to subtract the (always positive) x^2 contribution. But the earthbound twin snakes along the time direction with x close to zero, so her path must be the *longest possible path*. Actually, that is just another way of saying what we already know: that the earthbound twin is traveling as fast as possible through time and so it is she who ages the most.

Our explanation so far has been presented from the viewpoint of the earthbound twin. To fully satisfy ourselves that there is no paradox, we should see how things look from the viewpoint of the astronaut twin. For her, the earthbound twin is the one doing the traveling while she snakes along her own time axis. It looks like the paradox is back again; since the astronaut twin is at rest relative to her spaceship, it seems that she should speed maximally through time and hence age the most. But there is a very subtle point here. The distance equation does *not* apply if we set out to use the astronaut twin's clocks and rulers to measure distances and times. More precisely, it fails when the astronaut twin undergoes the acceleration that turns the spaceship around. Why does it fail? The arguments we presented when we figured it out seemed pretty watertight. But if one uses an accelerating system of clocks and rulers to make measurements, as the astronaut twin must, then the assumption that spacetime is unchanging and the same everywhere that we used to write down the distance equation is wrong. Over the time of the acceleration, the astronaut twin will be pushed back into her seat, in much the same way that you are pushed back into your seat when you press the accelerator pedal on a car. For a start, that immediately picks out a special direction in space: the direction of the acceleration. The existence of that force must be accounted for in the distance equation, and that is where the loophole resides. It is a little too complicated for us to go into the mathematical details, but the upshot is that when the spaceship fires its rockets to turn around, the earthbound twin ages rapidly relative to the astronaut twin and that more than makes up for the fact that she ages more slowly during the nonaccelerating phases of the expedition. There is no paradox.

We can't resist quoting some numbers, because the effect can be startling. Space travel is most comfortable for those onboard the spaceship if the rockets are firing in order to sustain an acceleration equal to "one g." That means that the space travelers feel their own weight inside the rocket. So let's imagine a journey of 10 years at that acceleration, followed by 10 more years decelerating at the same rate, at which point we turn the spaceship around and head back to Earth, accelerating for 10 more years and decelerating for a further 10 before finally arriving back. In total the travelers onboard the spaceship will have been journeying for a total of 40 years. The question is how many years have passed on Earth? We'll just quote the result because the mathematics is (only a little) beyond the level of this book. The result is that a breathtaking 59,000 years will have passed on Earth!

This has been a remarkable journey, and we hope the reader has followed us into the world of spacetime. We are now ready to head directly to $E = mc^2$. Armed with spacetime and our invariant definition of distance, we ask a simple but very important question: Are there other invariant quantities that also describe the properties of real objects in the real world? Of course, distances aren't the only things that are important. Objects have mass, they can be hard or soft, hot or cold, solid, liquid, or gas. Since all objects live in spacetime, is it possible to describe everything about the world in an invariant way? We will discover in the next chapter that it is, and the consequences are profound, for this is the road that leads directly to $E = mc^2$.

5

Why Does E=mc²?

In the last chapter we showed that merging space and time together into spacetime is a very good idea. Central to our whole investigation was the notion that distances in spacetime are invariant, which means that there is consensus throughout the universe as to the lengths of paths through spacetime. We might even regard it as a defining characteristic of spacetime. We were able to rediscover Einstein's theory but only if we interpreted the cosmic speed limit c as the speed of light. We haven't proved that c has anything to do with the speed of light yet, but we'll dig much more deeply into the meaning of c in this chapter. In a sense, however, we have already begun to demystify the speed of light. Because the speed of light appears in $E = mc^2$, it often seems as if light itself is important in the structure of the universe. But in the spacetime way of looking at things, light is not so special. In a subtle way, democracy is restored in the sense that everything hurtles through spacetime at the same speed, c, including you, planet Earth, the sun, and the distant galaxies. Light just happens to use up all of its spacetime speed quota on motion through space, and in so doing travels at the cosmic speed limit: The apparent specialness of light is an artifact of our human tendency to think of time and space as different things. There is in fact a reason why light is forced to use up its quota in this way, and this is intimately related to our goal of understanding $E = mc^2$.

$E = mc^2$ is an equation. As we have been at some pains to emphasize, to a physicist equations are a very convenient and powerful shorthand for expressing relationships between objects. In the case of $E = mc^2$ the "objects" are energy (E), mass (m), and the speed of light (c). More generally, the objects living inside an equation could represent real material things, such as waves or electrons, or they could represent more abstract notions—such things as energy, mass, and distances in spacetime. As we have seen previously in this book, physicists are very demanding of their fundamental equations, for they insist that everyone in the universe should agree upon them. This is quite a demand—and at some time in the future we might discover that it is not possible to hold on to this ideal. Such a turn of events would be quite shocking for any modern physicist, since the idea has proved astonishingly fruitful since the birth of modern science in the seventeenth century.

As good scientists, however, we must always acknowledge that nature has no qualms about shocking us, and reality is what it is. For now, all we can say is that the dream remains intact. We explored this ideal of universal agreement earlier in the book and expressed it very simply: The laws of physics should be expressed using invariant quantities. All of the fundamental equations of physics that we know today achieve this by being written in such a way that they express relationships between objects in spacetime. What exactly does that mean? What is an object that lives in spacetime? Well, anything that exists presumably exists in spacetime, and so when we come to write down an equation—for example, one that describes how an object interacts with its environment—then we should find a way to express this mathematically using invariant quantities. Only then will everyone in the universe agree.

A good example might be to consider the length of a piece of string. Based on what we have learned, we can see that although the piece of string is a meaningful object, we should avoid writing down an equation that deals only with its length in space. Rather, we should be more ambitious and talk about its length in spacetime, for that is the spacetime way. Of course, for earthbound physicists it might be convenient to use equations that express relationships between lengths in space and other such things—certainly engineers find that way of going about things very useful. The correct way to view an equation that uses only lengths in space or the time measured by a clock is that it is a valid approximation if we are dealing with objects that move very slowly relative to the cosmic speed limit, which is usually (but not always) true for everyday engineering problems. An example we have already met where this is not true is a particle accelerator, where subatomic particles whiz around in circles at very close to the speed of light, and live longer as a result. If the effects of Einstein's theory are not taken into account, particle accelerators simply stop working properly. Fundamental physics is all about the quest for fundamental equations, and that means working only with mathematical representations of objects that have a universal meaning in spacetime. The old view of space and time as distinct leads to a way of viewing the world that is something akin to trying to watch a stage play by looking only at the shadows cast by the spotlights onto the stage. The real business involves three-dimensional actors moving around and the shadows capture a two-dimensional projection of the play. With the arrival of the concept of spacetime, we are finally able to lift our eyes from the shadows.

All of this talk of objects in spacetime may sound rather abstract but there is a point to it. So far we have met one "mathematical representation of an object that has a universal meaning in spacetime"—the spacetime distance between two events. There are others.

Before we grapple with a new type of spacetime object we shall take one step back and introduce its analogue in the three dimensions of our everyday experience. It should come as no surprise (especially having read this far) that any reasonable attempt to describe the natural world exploits the notion of the distance between two points. Now, a distance is a special type of object—one that is characterized by a single number. For example, the distance from Manchester to London is 184 miles and the distance from the soles of your feet to the top of your head (more usually referred to as your height) is, at a guess, around 175 centimeters. The word following the number (cm or miles) just explains how we're doing the counting but in both cases a single number suffices. The distance from Manchester to London provides some useful information—enough to know how much fuel to put in your car, for example, but not quite enough to make the journey. Without a map we might well head off in the wrong direction and end up in Norwich.

A slightly surreal and very impractical solution to that problem would be to construct a giant arrow whose length is 184 miles. We could place one end of the arrow in Manchester and the tip could sit in London. Arrows are useful objects when physicists set about the business of describing the world: They capture simultaneously the idea that something can have a size and also a direction. Obviously our giant Manchester–London arrow makes sense only once it is placed in a particular orientation; otherwise we might still end up in Norwich. That is what we mean when we say that the arrow has both size and direction. The arrows used by weather forecasters to illustrate how the wind blows provide another example of how arrows can help us describe the world. The swirling arrows capture the essence of the flow of the wind, telling us in which direction it blows at any particular point on the map as well as the wind speed: The bigger the arrow, the stronger the wind. Physicists call objects that are represented by arrows vectors. The wind speed as demonstrated on the weather map and the giant Manchester–London arrow are vectors in two dimensions, needing only two numbers for their description. For example, we might say that the wind is blowing at 40 miles per hour in a southeasterly direction. By showing us arrows in only two dimensions, the weather forecasters are not giving us the whole story—they are not telling us if the air is moving upward or downward and by what degree, but that isn't something we are usually very interested in.

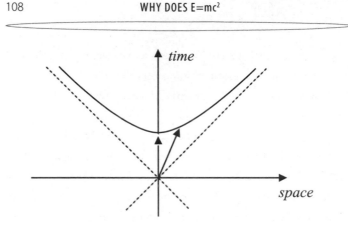

There can also be vectors in three or more dimensions. If we began our journey from Manchester to London in one of the old villages in the Pennine Hills north of Manchester, we would have to point our arrow slightly downward since London sits on the banks of the River Thames at sea level. Vectors living in the three dimensions of everyday space are described by three numbers. By now, you might have guessed that vectors can also exist in spacetime, and these will be described by four numbers.

We are now about to reveal the two remaining pieces on the road to $E = mc^2$. The first piece should come as no surprise— we are only ever going to be interested in vectors in the four dimensions of spacetime. That is easy to say but a weird concept: Just as a vector can point "north," we now have the notion of a vector that points "in the time direction." As is the norm when we talk about spacetime, this is not something we can picture in our mind's eye, but that is our problem, not nature's. The space-time landscape analogy of the last chapter might help you build a mental picture (at least of a simplified spacetime with only one dimension of space). Four-dimensional vectors will be characterized by four numbers. The archetypal vector is the one that connects two points in spacetime. Two examples are illustrated in Figure 9. That one of the vectors in Figure 9 points exactly in the time direction and that both just happen to start out from the same place is only for our convenience. Generally speaking, you should think of *any* two points in spacetime with an arrow joining them. Vectors like these are not entirely abstract things. Your going to bed at 10 p.m. and subsequent awakening at 8 a.m. defines an arrow linking two events in spacetime; it is "10 hours multiplied by c long" and it points entirely in the time direction. Moreover, we have actually been using vectors in spacetime throughout the book but haven't used the terminology before. For example, we met a very important vector in our discussion of the intrepid motorcyclist, journeying over the undulating landscape of spacetime with his throttle stuck. We worked out that the motorcyclist always travels at a speed c through spacetime, and the only choice he can make is the direction in which he points his motorcycle (although he doesn't even have complete freedom of direction, because he is restricted to staying within a bearing of 45 degrees of north). We can represent his motion with a vector of fixed length c, which points in the direction in which he is traveling over the spacetime landscape. This vector has a name. It is called the spacetime velocity vector. To use the correct terminology, we would say that the velocity vector always has length c and is restricted to point within the future lightcone. The lightcone is a fancy name for the area contained within the two 45-degree lines that are so important

in protecting causality. We can completely describe any vector
in spacetime by specifying how much of it points in the time direction and how much of it points in the space direction.

By now, we are familiar with the statement that the distances
in time and space between events are measured differently by
observers moving at different speeds relative to each other, but
they must change in such a way that the spacetime distance always remains the same. Because of the strange Minkowski
geometry, this means that the tip of the vector can move around
on a hyperbola that lies in the future lightcone. To be absolutely
concrete, if the two events are "going to bed at 10 p.m." and "waking up at 8 a.m.," then an observer in the bed concludes that the
spacetime distance vector points up his time axis, as illustrated
in Figure 9, and its length is simply the time elapsed on his
watch (10 hours) multiplied by c. Someone flying past at high
speed would be free to interpret the person in bed as doing the
moving. She would then have to add in a bit of space movement
as well when she viewed the person in bed, and that moves the
tip of the vector off her time axis. Because the arrow's length
cannot change, it must stay on the hyperbola. The second, tilted
arrow in Figure 9 illustrates the point. As you can see, the
amount of the vector pointing in the time direction has increased and this means that the fast-moving observer concludes
that more time passes between the two events (i.e., more than 10
hours elapses on her watch). This is yet another way to picture
the strange effect of time dilation.

So much, for now at least, for vectors (we will need the velocity spacetime vector again in a moment). The next few para-
graphs relate to the second crucial piece of the $E = mc^2$ jigsaw.

Imagine you are a physicist trying to figure out how the universe works. You are comfortable with the idea of vectors and on occasion you have written down mathematical equations that contain them. Now suppose that someone, perhaps a colleague, tells you there is a very special vector, one that has the property that it never changes, no matter what happens to that part of the universe to which it corresponds. Your first reaction might be to express disinterest—if nothing changes then it is hardly likely to be capturing the essence of the matter at hand. Your interest would probably perk up if your colleague told you that the single, special vector is built up by adding together a whole bunch of other vectors, each associated with a different part of the thing you are trying to understand. The various parts of the thing can jiggle around and, as they do so, each of the individual vectors can change, but always in such a way that the sum total of all the vectors adds up to the same unchanging special vector. Incidentally, adding vectors together is easy, and we shall return to it in a moment.

To illustrate just how useful this idea of unchanging vectors can be, let's think about a very simple task. We want to understand what happens when two billiard balls collide head-on. An example from billiards hardly sounds of earth-shattering significance but physicists quite often pick rather mundane examples like this, not because they can only study such simple phenomena or because they love billiards, but rather because concepts are often easiest to grasp first in simpler examples. Back to billiards: Your colleague explains that you should associate a vector with each ball. The vector should point in the direction of the ball's motion. The claim is that by adding together

FIGURE 10

the two vectors (one for each ball) we can obtain the special unchanging vector. That means that whatever happens in the collision, we can be sure that the two vectors associated with the balls after the collision will combine to make precisely the same vector as that obtained from the two balls before the collision. This is potentially a very valuable insight. The existence of the special vector severely limits the possible outcomes of the collision. We would be particularly impressed by our colleague's claim that the "conservation of these vectors" works for every system of things in the whole universe, from colliding billiard balls to the explosion of a star. It will probably come as no surprise to know that physicists don't go around referring to these as special vectors. Rather they speak of the momentum vector and the conservation of vectors is more commonly known as the conservation of momentum.

We have left a couple of points hanging: Just how long are the momentum arrows and exactly how are we to add them together? Adding them together is not hard; the rule is to place all of the arrows that we want to add together end-to-end. The net effect is to define an arrow that links the start of the first arrow in the chain to the tip of the last arrow. Figure 10 shows how it is done for three randomly chosen arrows. The big arrow is the sum of the little ones. The length of a momentum vector is something we can ascertain from experiments, and historically this is how it was arrived at. The concept itself dates back over a thousand years, simply because it is useful. In a crude sense, it expresses the difference between being hit by a tennis ball or an express train when both are traveling at 60 miles per hour. As we have discussed, it is closely related to the speed and, as the previous example illustrates all too vividly, it should also be related to mass. Pre-Einstein, a momentum vector has length that is simply the product of mass and speed. As we have already said, it points in the direction of motion. As an aside, the modern view of momentum as a quantity that is conserved relates to the work of Emmy Noether, as we discussed earlier. Then we learned of the deep connection between the law of conservation of momentum and translational invariance in space. In symbols, the size of the momentum of a particle of mass m moving with a speed v can be expressed as $p = mv$, where p is the commonly used symbol for momentum.

Up until now we have not really talked about what mass actually is, so before we proceed we ought to be a little more precise. An intuitive idea of mass might be that it is a measure of the amount of stuff something contains. Two bags of sugar have a mass twice that of one bag, and so on. Should we so desire, we could measure all masses in terms of the mass of a standard bag of sugar, using an old-fashioned set of balancing scales. This is how groceries used to be sold in shops. If you wanted to buy 1 kilogram of potatoes, you could balance the potatoes on a pair of scales against a kilogram bag of sugar, and everyone would accept that you had bought the right amount of potatoes.

Of course, "stuff" comes in lots of different types, so "amount of stuff" is horribly imprecise. Here is a better definition: We can measure mass by measuring weight. That is, heavier things have more mass. Is it that simple? Well, yes and no. Here on Earth, we can determine the mass of something by weighing it, and that is what everyday bathroom scales do. Everyone is familiar with the idea that we "weigh" in kilograms and grams (or pounds and ounces). Scientists would not agree with that. The confusion arises because mass and weight are proportional to each other if you measure them close to the surface of the earth. You might like to ponder what would happen if you took your bathroom scales to the moon. You would in fact weigh just over six times less than you do on Earth. You really do weigh less on the moon, but your mass has not changed. What has changed is the exchange rate between mass and weight, although twice the mass will have twice the weight wherever it is measured (we say that weight is proportional to mass).

Another way to define mass comes from noticing that more massive things take more pushing to get them moving. This feature of nature was expressed mathematically in the second most famous equation in physics (after $E = mc^2$, of course): $F = ma$, first published in 1687 by Isaac Newton in his *Principia Mathematica*. Newton's law simply says that if you push something with a force F, that thing starts to accelerate with an acceleration a. The m stands for mass, and you can therefore work out how massive something is experimentally by measuring how much force you have to apply to it to cause a given acceleration. This is as good a definition as any, so we'll stick with it for now. Although if you have a critical mind you might be worrying as to how exactly we should define "force." That is a good point but we won't go into it. Instead we will assume that we know how to measure the amount of push or pull, a.k.a. force.

That was a fairly extensive detour, and while we haven't really said what mass is at a deep level, we've given the "school textbook" version. A deeper view as to the very origin of mass will be the subject of Chapter 7, but for now it is presumed to "just be there"—an innate property of things. What is important here is that we are going to assume that mass is an intrinsic property of an object. That is, there should be a quantity in spacetime that everyone agrees upon called mass. This should therefore be one of our invariant quantities. We haven't advanced any argument to convince the reader that this quantity necessarily should be the same as the mass in Newton's equation, but as with many of our assumptions, the validity or otherwise will be tested when we have derived the consequences. We will now return to billiards.

If the two balls collide head-on, and they have the same mass and the same speed, then their momentum vectors are equal in length but point in opposite directions. Add them together and the two cancel each other entirely. After the collision, the law of momentum conservation predicts that whatever the particles will be doing, they must come off with equal speeds and in opposite directions. If this were not the case, then the net momentum afterward could not possibly cancel out. The law of momentum conservation is, as we said, not confined to billiard balls. It works everywhere in the universe, and that is why it is so very important. The recoil of a cannon after it shoots a cannonball or the way in which an explosion sprays particles in every direction are both in accord with momentum conservation. Actually, the case of the cannonball is worth a little more of our attention.

Before the cannon is fired, there is no net momentum and the cannonball is sitting at rest inside the barrel of the cannon, which is itself standing still on top of a castle. When the cannon is fired, the cannonball shoots out at high speed, while the cannon itself recoils a bit but stays pretty much where it began, fortunately for the soldiers in the castle who fired it. The cannonball's momentum is specified by its momentum vector, which is an arrow whose length is equal to the mass of the ball multiplied by its speed and whose direction points away from the cannon along the direction of flight as it emerges from the barrel. Momentum conservation tells us that the cannon itself must recoil with a momentum arrow that is exactly equal in length but opposite in direction to the arrow associated with the ball. But since the cannon is much heavier than the ball, the cannon recoils with much less speed. The heavier the cannon, the slower it recoils. So, big and slow things can have the same momentum as small and fast ones. Of course, both the cannon and the ball slow down eventually (and lose momentum as a result), and the ball changes its momentum because it is acted on by gravity. However, this does not mean that momentum conservation has gone wrong. If we could take account of the momentum taken by the air molecules that collide with the ball and the molecules inside the bearings of the cannon, and the fact that the momentum of the earth itself changes slightly as it interacts with the ball through gravity, then we would find that the total momentum of everything would be conserved. Physicists usually cannot keep track of where all of the momentum is going when things like friction and air resistance are present, and as a result the law of momentum conservation is usually applied only when external influences are not important. It is a slight weakening of the scope of the law, but it ought not to detract from its significance as a fundamental law of physics. That said, let's see if we can finish our game of billiards, which is dragging on somewhat.

To simplify matters, imagine that frictional forces are completely removed so that all we have to think about are the colliding billiard balls. Our newfound law of momentum conservation is very valuable but it isn't a panacea. It isn't in fact possible for us to figure out the speed of the billiard balls after their collision knowing only that momentum is conserved and the masses and velocities of the balls before the collision. To be able to work this out, we need to make use of another very important conservation law.

We have introduced the ideas that moving things can be described by a momentum vector and that the sum of all momentum vectors remains constant for all time. Momentum is interesting to physicists precisely because it is conserved. It is important to be clear on this fact. If you don't like the word "momentum," then you could do much worse than to speak of "the arrow that is conserved." Conserved quantities are, as we are beginning to discover, rather numerous and exceedingly useful in physics. Generally speaking, the more conservation laws you have at your disposal when tackling a problem, the easier it will be to find a solution. Of all the conservation laws, one stands out more than any other, because of its profound usefulness. Engineers, physicists and chemists uncovered it very slowly during the course of the seventeenth, eighteenth, and nineteenth centuries. We are speaking of the law of conservation of energy.

In the first instance, energy is an easier concept to grasp than momentum. Like momentum, things can have energy but, unlike momentum, energy has no direction. In that respect it is more like temperature, in that a single number will suffice to specify it. But what is "energy"? How do we define it? What is it measuring? Momentum was easy in that regard: An arrow points in the direction of motion and is of a length equal to the product of the mass and the speed. Energy is less easy to pin down, because it can come in many different guises, but the bottom line is clear enough: Whatever happens, the sum total of all the energy in any process should remain unchanged regardless of how things might be changing. Again, Noether gave us the deep explanation. The conservation of energy arises because the laws of physics remain unchanged with time. That statement does not mean that things do not happen, which would obviously be silly. Instead it means that if Maxwell's equations hold true today, then they ought also to hold true tomorrow. You can replace "Maxwell's equations" with any fundamental law of physics—Einstein's postulates, for example.

That said, and as with the conservation of momentum, the conservation of energy was first discovered experimentally. The story of its discovery is a meander though the history of the Industrial Revolution. It sprang from the work of many a practical experimenter who came across an immense variety of mechanical and chemical phenomena in pursuit of industrial Jerusalem. Men like the unfortunate Count Rumford of Bavaria (born Benjamin Thompson in Massachusetts in 1753), whose job it was to bore cannon for the Duke of Bavaria. While boring away, he noticed that the metal of the cannon and the drill bit got hot, and correctly surmised that the rotational motion of the drill was being converted into heat by friction. This is the opposite of what happens in a steam engine, in which heat gets converted into the rotary motion of the wheels of a train. It seemed natural to associate some common quantity with heat and rotational motion, since these seemingly different things appear interchangeable. This quantity is energy. Rumford has been termed unfortunate because he married the widow of another great scientist, Antoine Lavoisier, after Lavoisier lost his head to the guillotine in the French Revolution, in the mistaken belief that she would do for him as she had for Lavoisier and dutifully take notes and obey him as a good eighteenth-century wife should. It turned out that she had been submissive only under the duress of Lavoisier's iron will, and in his rather wonderful book *The Quest for Absolute Zero*, Kurt Mendelssohn described her as leading him "a hell of a life" (the book was written in 1966, hence the quaint turn of phrase). The key point is that energy is always conserved, and it is because it is conserved that it is interesting.

Ask someone on the street to explain what energy is and you'll get either a sensible answer or a pile of steaming New Age nonsense. There is such a wide spectrum of meanings in popular culture because "energy" is a word that is widely used. For the record, energy has a very precise definition indeed and it cannot be used to explain ley lines,* crystal healing, life after death, or reincarnation. A more sensible person might answer that energy can be stored away, inside a battery waiting in suspension until someone "completes the circuit"; it could be a measure of the amount of motion, with faster objects having

---

* Supposed points on Earth that resonate "psychic energy."

more energy than slower ones. Energy stored in the sea or in the wind provide particular examples of that. Or perhaps you would be told that hotter things contain more energy than colder ones. A giant flywheel inside a power station can store up energy, to be released onto the national electrical grid to meet the demands of an energy-hungry population, and energy can be liberated from inside an atomic nucleus to generate nuclear power. These are just some of the ways we might encounter energy in everyday life, and they can all be quantified by physicists and used to balance the books when it comes to making sure that the net effect of any process is such that the total energy remains unchanged.

To see energy conservation in action in a simple system, let us return to the colliding billiard balls for the final time. Before they hit each other, each ball has some energy due to its motion. Physicists call that type of energy kinetic energy. The *Oxford English Dictionary* defines the word "kinetic" to mean "due to or resulting from motion," so the name is sensible. We previously assumed that the balls were traveling at equal speeds and had the same mass. They then collide and head out at equal speeds and in opposite directions. That much is dictated by momentum conservation. Closer inspection reveals that their outgoing speed is a little less than the speed before the impact. That is because some of the initial energy has been dissipated in the collision. The most apparent dissipation occurs with the emission of sound. As the balls collide, they agitate the molecules in the surrounding air, and this disturbance makes its way to our ears. So some of the initial energy leaks away, leaving less for the outgoing billiard balls. As far as our journey in this book is concerned, we don't actually need to know how to quantify energy in all of its different guises, although the formula for kinetic energy will turn out to be useful later. To anyone who has a little experience in high school science, it will be indelibly imprinted deep within their psyche: kinetic energy $= \frac{1}{2}mv^2$. The main thing is to realize that energy can be quantified in a single number and, provided we are careful with the bookkeeping, the total energy in a system remains constant for all time.

Now let us get back to the point. We introduced momentum as an example of a quantity that is described by an arrow and, along with energy, its utility arises out of the fact that it is a conserved quantity. That all seems well and good but a huge dilemma is lurking in the shadows. Momentum is an arrow that lives only in the three dimensions of our everyday experiences. Generally speaking, a momentum arrow can point up or down or southeast or in any other direction in space. This is because things can and do fly around in any direction in space, and the momentum arrow captures the direction of motion. But the whole point of the last chapter was to expose our tendency to isolate space and time as a fallacy. We need arrows that point in the four dimensions of spacetime; otherwise, we'll never be able to build fundamental equations that respect Einstein. To reiterate: Fundamental equations should be built out of objects that live in spacetime, not objects that live in space or in time separately because those types of object are subjective. Recall that neither the length of an object in space nor the time interval between two events are quantities whose values everyone will agree upon. That is what we mean when we say they are subjective. Likewise, momentum is an arrow that points somewhere

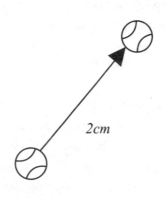

only in space. That bias against time sows the seeds of its destruction. Does spacetime herald the breakdown of this most fundamental of laws in physics? It is true that our newly discovered structure of spacetime sows the seeds of destruction but it also indicates how we should proceed: We need to find an invariant quantity to replace the old three-dimensional momentum. This is a key point in our narrative: Such a thing does exist.

Let's take a closer look at the three-dimensional momentum vector. Figure 11 shows an arrow in space. It might represent the amount by which a ball moves as it rolls across a table.* To be more precise, suppose that at midday the ball is at one end of the arrow, then 2 seconds later it is at the other end, the tip. If the ball moves 1 centimeter each second, then the arrow is 2 centimeters long. The momentum vector is easy to obtain. It is an

---

* There is nothing special about it being a ball; it could be any object.

arrow pointing in exactly the same direction as the arrow in Figure 11 except that its length is different. The length is equal to the speed of our ball (in this case 1 centimeter per second) multiplied by the mass of the ball, which we might suppose to be 10 grams. Physicists would say that the momentum vector of the ball has a length of 10 gram-centimeters per second (which they would abbreviate to something like 10 g cm/s). It is again going to be well worth our while to be a little bit more abstract and introduce placeholders rather than commit to any particular mass or speed. As ever, we certainly do not wish to transmogrify into the school mathematics teachers of our youth. But . . . if Δx is a placeholder for the length of the arrow, Δt is the time interval, and m is the mass of the ball ($\Delta x = 2$ centimeters, $\Delta t = 2$ seconds, and $m = 10$ grams in the example), then the momentum vector has a length equal to $m\Delta x/\Delta t$. It is common in physics to use the Greek symbol Δ (pronounced "delta") to represent "difference," and in that spirit Δt stands for the difference in time or the time interval between two things, and Δx stands for the length of something, in this case the distance in space between the start and the end of our measurement of the ball's position.

We have succeeded in constructing the momentum vector of a ball in three-dimensional space, although it is hardly the most exciting thing we have done. We're now going to make the bold step of trying to build a momentum vector in spacetime, and we will do it in an entirely analogous way to the three-dimensional case. The only constraint is that we will use only objects that are universal in spacetime.

Again we shall start with an arrow, this time pointing in four-dimensional spacetime, as illustrated in Figure 12. One end of

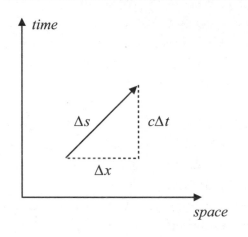

FIGURE 12

the arrow specifies where our ball is at one instant and the other end specifies where it is some time later. The length of the arrow must be determined by Minkowski's formula for the distance in spacetime, and it is therefore specified by $(\Delta s)^2 = (c\Delta t)^2 - (\Delta x)^2$. Remember that Δs is the only length that everyone in the universe can agree upon (something that most definitely cannot be said for Δx and Δt separately), and as such it is the distance measurement we must use, taking the place of Δx in the three-dimensional definition of momentum. But what is to take the place of the time interval Δt? (Remember, we are trying to find a four-dimensional replacement for $m\Delta x/\Delta t$). Here comes the crunch: We cannot use Δt because it is not a spacetime invariant. Not everyone agrees on time intervals, as we have emphasized again and again, and therefore we must not use time intervals in our quest for the four-dimensional momentum. What are our choices? By what could we possibly divide the length of the arrow by to determine the ball's speed through spacetime?

We want to construct something that is an improvement over the old three-dimensional momentum. If we are dealing with objects moving around at speeds that are slow compared to the speed of light, then we should find that the new momentum is at least approximately equivalent to the old one. If that is to happen, we must divide the length of our arrow in spacetime Δs by some quantity that is of the same type as an interval in time. Otherwise the new four-dimensional momentum will be an entirely different beast from the old three-dimensional momentum. Intervals of time can be measured in seconds, so we would also like something that can be measured in seconds. Starting from our invariant spacetime quantities, the speed of light c and the distance Δs, there is only one viable combination: It is the number we obtain upon dividing the length of the arrow (Δs) by the speed c. In other words, if Δs is measured in meters, and the speed c is measured in meters per second, then $\Delta s/c$ is measured in seconds. This must be the number we need to divide the length of our arrow by, since it is the only invariant thing we have at our disposal that is measured in the correct currency. So let us go ahead and divide Δs by the time $\Delta s/c$. The answer is simply c (for much the same reason that 1 divided by ½ is equal to 2). In other words, the four-dimensional analogue of the speed in our three-dimensional momentum formula is the universal speed limit c.

This all might feel rather familiar, and that is because it should be familiar. All we have done is to calculate the speed of an object (a ball in our example) in spacetime and found it to be *c*. We came to exactly the same conclusion in the previous chapter when we considered the motorcyclist moving over the spacetime landscape. From the perspective of this chapter, we have done rather more because we have also found a spacetime velocity vector that has the potential to be used in a new definition of four-dimensional momentum. The velocity of an object moving through spacetime always has length *c* and it points in the direction in spacetime in which the object travels.

To finish our construction of the new spacetime momentum arrow, all we need to do is multiply the spacetime velocity vector by the mass *m*. It follows that our proposed momentum arrow always has a length equal to *mc* and points in the direction of travel of the object in spacetime. At first glance this new momentum arrow is a little boring because its length in spacetime is always the same. It seems we are hardly off to a good start. But we should not be deterred. It remains to be seen whether the spacetime momentum vector that we have just constructed bears any relation to the old-fashioned three-dimensional momentum or, for that matter, whether it will be of any use to us in our new spacetime world.

To delve a little deeper, we will now take a look at the portions of our new spacetime momentum vector that point in the space and time directions separately. To do this bit of delving, we need a bit of absolutely unavoidable mathematics. We can only apologize to the nonmathematical reader and promise that we will go very slowly. Remember, it is always an option to skim over the equations in search of the punch line. The mathematics makes the argument more convincing but it is okay to read on without following the details. Similarly, we must also apologize to the reader familiar with mathematics for laboring the point. We have a saying in Manchester: "You can't have your cake and eat it." This saying is perhaps harder to understand than the mathematics.

Recall that we arrived at an expression for the length of the momentum vector in three-dimensional space, $m\Delta x/\Delta t$. We have just argued that Δx should be replaced by Δs and Δt should be replaced by $\Delta s/c$ to form the four-dimensional momentum vector, which has a seemingly rather uninteresting length of mc. Indulge us for one more paragraph, and let us write the replacement for Δt, i.e., $\Delta s/c$, in full. $\Delta s/c$ is equal to $\sqrt{(c\Delta t)^2 - (\Delta x)^2}/c$. This is a bit of a mouthful, but a little mathematical manipulation allows us to write it in a simpler form, i.e., it can also be written as $\Delta t/\gamma$ where $\gamma = 1/\sqrt{1 - v^2/c^2}$. To obtain that, we have used the fact that $v = \Delta x/\Delta t$ is the speed of the object. Now γ is none other than the quantity we met in Chapter 3 that quantifies the amount by which time slows down from the point of view of someone observing a clock fly past at speed.

We are actually nearly where we want to be. The whole point of that piece of mathematics is that it allows us to figure out by exactly how much the momentum vector points off in the space and time directions separately. First let's recap how we dealt with the momentum vector in three-dimensional space. Figure 11 helped us picture this. The three-dimensional momentum vector points off in exactly the same direction as the arrow in Figure 11, because it points in the same direction that the ball is moving in. The only difference is that its length is changed because we need to multiply it by the mass of the ball and divide by the time interval. The situation is entirely analogous in the four-dimensional case. Now the momentum vector points off in the direction in spacetime in which the ball is moving, which is the direction of the arrow in Figure 12. Again, to get the momentum, we need to rescale the length of the arrow, but this time we are to multiply by the mass and divide by the invariant quantity $\Delta s/c$ (which we showed in the last paragraph is equal to $\Delta t/\gamma$). If you look carefully at the arrow in Figure 12, you should be able to see that if we want to change the length by some amount while keeping it pointing in the same direction, then we must simply change the bit pointing in the x direction (Δx) and the bit pointing in the time direction ($c\Delta t$) by the same amount. So, the length of the part of the momentum vector that points in the space direction is simply Δx multiplied by m and divided by $\Delta t/\gamma$, which can be written as $\gamma m\Delta x/\Delta t$. Remembering that $v = \Delta x/\Delta t$ is the speed of the object through space, we have the answer: The part of the momentum spacetime vector that points in the space direction has a length equal to γmv.

Now that really is interesting—the momentum vector in spacetime that we just constructed is not boring at all. If the speed v of our object is much less than the speed of light c, then γ is very close to one. In that case, we regain the old-fashioned momentum, namely the product of the mass with the speed $p = mv$. This is very encouraging—we should press on. In fact, we have done much more than translate the old-fashioned momentum into the new four-dimensional framework. For one thing, we have what is presumably a more accurate formula since γ is only ever exactly one when the speed is zero.

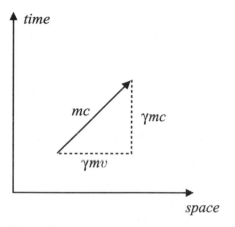

FIGURE 13

More interesting than the fact that we have modified $p = mv$ is what happens when we consider that part of the momentum vector that points off in the time direction. After all of the hard work we have been investing, it is not hard for us to compute it, and Figure 13 shows the answer. That part of the new momentum vector that points off in the time direction has a length equal to $c\Delta t$ multiplied by m and divided by $\Delta t/\gamma$ again, which is γmc.

Remember, momentum is interesting to us because it is conserved. Our goal has been to find a new, four-dimensional momentum that will be conserved in spacetime. We can imagine a bunch of momentum vectors in spacetime, all pointing off in different directions. They might, for example, represent the momenta of some particles that are about to collide. After the collision, there will be a new set of momentum vectors, pointing in different directions. But the law of momentum conservation tells us that the sum total of all the new arrows must be exactly the same as the sum total of the original arrows. This in turn means that the sum total of the portions of each of the arrows pointing in the space direction must be conserved, as should the sum of the portions pointing in the time direction. So if we tally up the values of γmv for each particle, then the grand total before the collision should be the same as the value afterward. Likewise for the time portions, but this time it is the sum total of the γmc values that is conserved. We appear to have two new laws of physics: γmv and γmc are conserved quantities. But what do these two particular things correspond to? At first sight, there is nothing much to get excited about. If speeds are small, then γ is very close to 1 and γmv simply becomes mv. We have therefore regained the old-fashioned law for momentum conservation. This is reassuring since we hoped that we would arrive at something that Victorian physicists would recognize. Brunel and the other great engineers of the nineteenth century certainly managed just fine without spacetime, so our new definition of momentum really had to give rise to almost the same answers as it did during the Industrial Revolution, provided things are not whizzing around at too close to the speed of light. After all, the Clifton Suspension Bridge did not suddenly cease to remain suspended when Einstein came up with relativity.

What can we say about the conservation of γmc? Since c is a universal constant upon which everyone always agrees, then the conservation of γmc is tantamount to saying that mass is conserved. That doesn't seem a big surprise and it is in accord with our intuition, although it is rather interesting that it has popped out as if from nowhere. For example, it seems to say that after burning coal in a fire, the mass of the ashes afterward (plus the mass of any matter that went up the chimney) should be equal to the mass of the coal before the fire was lit. The fact that γ isn't exactly one hardly seems to matter, and we might be tempted to move on, satisfied that we have already achieved a great deal. We have defined momentum in such a way that it is a meaningful quantity in spacetime and as a result we have derived (usually tiny) corrections to the nineteenth-century definition of momentum while simultaneously deriving the law of conservation of mass. What more could we hope for?

It has taken us a long time to reach this point, but there is a sting in the tail of this narrative. We are going to take a closer look at that part of the momentum vector that points off in the time direction, and in so doing we will, almost miraculously, uncover Einstein's most famous formula. The finale is within sight. Thales of Miletus is reclining in his bath, preparing for the ultimate enchantment. In following the book up to this point, you may well be juggling a lot of mental balls as you read this sentence. It is no mean feat, because you have learned a great deal of what a professional physicist might be expected to know about four-dimensional vectors and Minkowski spacetime. We are now ready for the climax.

We have established that γmc should be conserved. We need to be clear on what that means. If you imagine a game of relativistic billiards, then each ball has its own value for γmc. Add all those values up and whatever the total is, it does not change. Now let us play what at first seems a rather pointless game. If γmc is conserved, then so too is γmc^2, simply because c is a constant. Why we did that will become clear shortly. Now, γ is not exactly equal to one, and for small speeds it can actually be approximated by the formula $\gamma = 1 + \frac{1}{2}(v^2/c^2)$. You can check for yourself, using a calculator, that this formula works pretty well for speeds that are small compared to c^*. Hopefully the table below will convince you if you don't have a calculator handy. Notice that the approximate formula (which generates the numbers in the third column) is actually very accurate even for speeds as high as 10 percent of the speed of light ($v/c = 0.1$), which is a usually impossible-to-reach 30 million meters per second.

After making this simplification, γmc^2 is then approximately equal to $mc^2 + \frac{1}{2}mv^2$. It is at this point that we are able to realize the profoundly significant consequences of what we have been doing. For speeds that are small compared to c, we have determined that the quantity $mc^2 + \frac{1}{2}mv^2$ is conserved. More precisely, it is the quantity γmc^2 that is conserved, but at this stage, the former equation is much more illuminating. Why? Well, as we have already seen, the product $\frac{1}{2}mv^2$ is the kinetic energy we encountered in our example of the colliding billiard balls and it measures how much energy an object of mass m has

v/c	γ	$1 + \frac{1}{2}(v^2/c^2)$
0.01	1.00005	1.00005
0.1	1.00504	1.00500
0.2	1.02062	1.02000
0.5	1.15470	1.12500

TABLE 5.1

---

* i.e., it gives almost the same value as the exact formula: $\gamma = 1/\sqrt{1 - v^2/c^2}$.

as a result of the fact that it is moving with a speed v. We have discovered that there is a thing that is conserved that is equal to something (mc^2) plus the kinetic energy. It makes sense to refer to the "something that is conserved" as the energy, but now it has two bits to it. One is $\frac{1}{2}mv^2$ and the other is mc^2. Don't be confused by the fact that we multiplied by c. We did that only so our final answer included the term $\frac{1}{2}mv^2$ rather than $\frac{1}{2}mv^2/c^2$, and the former is what scientists have for many generations called kinetic energy. If you like, you can christen $\frac{1}{2}mv^2/c^2$ the "kinetic mass" or any other name you care to dream up. The name is irrelevant (even if it carries the great gravitas that "energy" does). All that matters is that it is the "time component of the momentum spacetime vector," and that it is a conserved quantity. Admittedly, the equation "the time component of the momentum spacetime vector equals mc" does not have the catchy appeal of $E = mc^2$, but the physics is the same.

Remarkably, we have demonstrated that the conservation of momentum in spacetime leads not only to a new, improved version of the conservation of momentum in three dimensions, but also to a revised law for the conservation of energy. If we imagine a system of particles all jiggling about, then we have just figured out that adding together the kinetic energy of all the particles plus the mass of all the particles multiplied by c squared we get something that is unchanging. Now, the Victorians would have been happy with the assertion that the sum of kinetic energies should be unchanging, and they would also have been happy with the assertion that the sum of the masses should be unchanging (multiplying by c squared is irrelevant when we're thinking about what is unchanging). Our new law is consistent with that being the case, but it is much more than that. As it stands there is nothing at all preventing some of the mass from being converted into kinetic energy and vice versa, as long as the sum of these two things is always conserved. We have discovered that mass and energy are potentially inter-changeable and the amount of energy we can extract from a mass m at rest (γ is equal to one in that case) is captured by the equation $E = mc^2$.

Our friend Thales of Miletus can at last achieve complete enchantment. He rises from his bath, dripping asses' milk onto the floor, and welcomes his concubines into his magnificent presence.

Let's recap: We wanted to look for an object in spacetime that did the job of momentum in three-dimensional space, because momentum is a conserved quantity and therefore useful. We were able to find such an object by building it only out of things that everyone agrees upon, namely the distance in spacetime, the universal speed limit, and the mass. The spacetime momentum vector that we constructed turned out to be very interesting. By looking at the part that points along the space direction, we re-discovered the old law of momentum conservation, with a tweak for things moving close to the speed of light. But the real gold came from looking at the part of the vector that points along the time direction. This gave us an entirely new version of the law of conservation of energy. The old-fashioned kinetic energy, $\frac{1}{2}mv^2$, was there, but a totally new piece appeared: mc^2. Thus, even if an object is standing still, it has energy associated with it, and that energy is given by Einstein's famous equation: $E = mc^2$.

What does it all mean? We have established that energy is an interesting quantity because it is conserved: "You can increase energy over here provided you lower it over there." Moreover, we have established that the raw mass of an object provides a potential source of energy. We can imagine taking a blob of matter, say 1 kilogram of "stuff" (it doesn't matter what) and "doing something to it" so that afterward there is no 1 kilogram of stuff anymore. And by that we don't mean the 1 kilogram has been smashed up into tiny bits, we mean that it has vanished. In fact, we can imagine an extreme scenario where all of the original mass gets used up. In its place must be 1 kilogram worth of energy (plus any energy we might have put in when we did the "doing something to it"). That energy could itself be in the form of mass, for example a few hundred grams of new "stuff" might be created, and the remaining energy could be in the form of kinetic energy: the new stuff could be whizzing about with speed. Of course, we just made all of that up; it was an imaginary scenario. The point to appreciate is that this is the kind of thing that *could* be allowed by Einstein's theory. Before Einstein, no one had dreamed that mass could be destroyed and converted into energy because mass and energy seemed to be entirely disconnected entities. After Einstein, everyone had to accept that they are different manifestations of the same type of thing. This is because we have discovered that energy, mass, and momentum must all be combined into a single spacetime object that we have been referring to as the spacetime momentum vector. Actually, its more usual name in physics circles is the energy-momentum four-vector. Just as we discovered that space and time should no longer be thought of as separate entities, so we have found that energy and momentum are shadows of a more profound object, the energy-momentum four-vector. We are fooled into thinking of them as unrelated and distinct entities because of our heavy intuitive bias to separate space and time from each other. Crucially, nature does exploit the opportunity— it is possible to convert mass into energy. If nature did not allow this to happen, then we would not even exist.

Before we unpick that rather strong statement, a further word on what we mean by "destroyed" is probably in order. We do not mean destruction in the sense that a precious vase might fall and get smashed into smithereens. After that kind of destruction you could imagine dejectedly sweeping up the pieces and weighing them—there would be no noticeable change in mass. What we mean is that the vase gets destroyed such that after the act of destruction there are fewer atoms than before and the mass is correspondingly less. This might seem like a new and controversial notion. The idea that matter is made up of tiny pieces and that we can chop the pieces up and rearrange them but never destroy them is a powerful one, dating back to Democritus in ancient Greece. Einstein's theory overturns that view of the world and leads instead to a world in which matter is more nebulous—capable of popping into and out of existence. Indeed, that cycle of destruction and creation is today carried out routinely in the world's particle physics accelerators. We shall come back to these matters later.

Now for the grand finale. Unfortunately, we have run out of things for Thales to do in polite company, but this is really going to be wonderful. We want to wrap up the identification of c with the speed of light. As we have been keen to stress, the important thing in the spacetime way of thinking about things is that c is a universal cosmic speed limit, not that it is the speed of light. In the last chapter we did eventually identify c as the speed of light but only after comparing to the results we found in Chapter 3. Now we can do it without resorting to ideas outside of the spacetime framework. We shall attempt to find an alternative interpretation of the c that occurs in $E = mc^2$, other than that it is the cosmic speed limit.

The answer can be found in another bizarre and well-hidden feature of Einstein's mass-energy equation. To investigate further, we need to step back from our approximations and write the space and time parts of the energy-momentum four-vector in their exact form. The energy of an object, which is the time part of the energy-momentum four-vector (multiplied by c), is equal to γmc^2, and the momentum, which is the space part of the energy-momentum four-vector, is γmv. Now we ask what at first sight seems to be a very weird question: What happens if an object has zero mass? A quick glance might suggest that if the mass is zero, then the object always has zero energy and zero momentum, in which case it would never influence anything and it might as well not exist. But thanks to a mathematical subtlety that is not the case. The subtlety lies in γ. Recall that $\gamma = 1/\sqrt{1 - v^2/c^2}$. If the object moves at the speed c, then the factor γ becomes infinite, because we have to take one divided by zero (the square root of zero is zero). So we have a strange situation for the very specific case in which the mass is zero and the speed is c. In the mathematical expressions for both momentum and energy, we end up with infinity multiplied by zero, which is mathematically undefined. In other words, the equations as they stand are use-less but, crucially, we are not entitled to conclude that the energy and momentum are necessarily zero for massless particles. We can, however, ask what happens to the ratio of the momentum and the energy. Dividing $E = \gamma m c^2$ by $p = \gamma m v$ leaves us with $E/p = c^2/v$, which for the special case $v = c$ leaves us with the equation $E = cp$, which is meaningful. Therefore, the bottom line is that both the energy and momentum could conceivably be nonzero even for an object with zero mass but *only* if that object travels at speed c. So Einstein's theory allows for the possible existence of massless particles. Here is where the experiments come in handy. They have shown us that light is made up of particles called photons and that as far as anyone can tell they have zero mass. As a result, they must travel at the speed c. There is an important point here—if at some point in the future an experiment is performed that reveals that photons actually have a tiny mass, what should we do? Well, hopefully you can answer that question now. The answer is that we do nothing, except go back to Einstein's second postulate in Chapter 3 and replace it with the statement that "the speed of massless particles is a universal constant." Certainly c remains unchanged by the new experimental data; what changes is that we should no longer identify it with the speed at which light travels.

This is pretty profound stuff. The c in $E = mc^2$ has something to do with light only because of the experimental fact that particles of light just happen to be massless. Historically, this was incredibly important because it allowed experimentalists like Faraday and theorists like Maxwell to gain direct access to a phenomenon that traveled at the special universal speed limit—electromagnetic waves. This played a key role in Einstein's thinking, and perhaps without this coincidence, Einstein would not have discovered relativity. We shall never know. "Coincidence" may be the right word because, as we shall see in Chapter 7, there is no fundamental reason in particle physics that guarantees that the photon should be massless. Moreover, there is a mechanism known as the Higgs mechanism that could, in a different universe, perhaps, have given it a nonzero mass. The c in $E = mc^2$ should therefore be seen more correctly as the speed of massless particles, which are absolutely forced to fly around the universe at this speed. From the spacetime perspective, c was introduced so we could define how to compute distances in the time direction. As such, it is ingrained into the very fabric of spacetime.

It may not have escaped your attention that the energy associated with a certain mass carries with it a factor of the speed of light squared. Since the speed of light is so great compared to everyday, run-of-the-mill speeds (the v in $\frac{1}{2}mv^2$) it ought to come as no surprise that the energy locked away inside even quite small masses is mind-bogglingly large. We are not yet claiming to have proven that this energy can be accessed directly. But if we could get at it, then how huge an energy supply could we be, quite literally, sitting on? We can even put a number on it because we have the relevant formulas on hand. We know that the kinetic energy of a particle of mass m moving with a speed v is approximately equal to $\frac{1}{2}mv^2$ and the energy stored up inside the mass is equal to mc^2 (we shall assume that v is small compared to c; otherwise, we would need to use the more complicated formula γmc^2). Let's play around with some numbers to get a better feel for what these equations actually mean.

A lightbulb typically radiates 100 joules of energy every second. A joule is a unit of energy named after James Joule, one of the great figures of Manchester whose intellectual drive powered the Industrial Revolution. One hundred joules every second is 100 watts, named after the Scottish engineer James Watt. The nineteenth century was a century of fantastic progress in science, now commemorated in the way we measure everyday quantities. If a city has 100,000 inhabitants, then a reasonable estimate is that it needs an electrical power supply of around 100 million watts (100 megawatts). To generate even 100 joules of energy requires a fair amount of mechanical effort. It is approximately equal to the kinetic energy of a tennis ball traveling at around 135 miles per hour, which is the service speed of a professional tennis player. You can go ahead and check this number. The mass of a tennis ball is around 57 grams (or 0.057 kilograms) and 135 miles per hour is nearly the same as 60 meters per second. If we put these numbers into $\frac{1}{2}mv^2$, we get a kinetic energy equal to ½ x 0.057 x 60 x 60 joules. One joule can be defined as the kinetic energy of a 2-kilogram mass traveling at 1 meter per second (that is why we converted the speed from miles per hour to meters per second), and you can do the multiplication yourself. One would therefore require a constant barrage of such tennis balls (one every second) to power just one electric lightbulb. In reality, the balls would have to travel even faster or arrive even more frequently because we would need to extract the kinetic energy from the balls, convert it to electrical energy (via a generator), and deliver it to the lightbulb. That is certainly a lot of effort to power a lightbulb.

How much mass would we need to do the same job if we could exploit Einstein's theory and convert it all into energy? Well, the answer is that the mass should equal the energy divided by the speed of light squared: 100 joules divided by 300 million meters per second, twice. This is just over 0.000000000001 grams or, in words, one-millionth of one-millionth (i.e., one-trillionth) of 1 gram. At that rate, we need to destroy only 1 microgram of material every second to power a city. There are around 3 billion seconds in one century, so we would need only 3 kilograms of material to keep the city going for 100 years. One thing is for sure, the energy potential that is locked away within matter is on a different scale from anything we ordinarily experience, and if we could unlock it, we would have solved all of the earth's energy problems.

Let us make one final point before we move on. The energy locked up in mass feels utterly astronomical to us here on Earth. It is tempting to say that this is because the speed of light is a very big number, but that is to emphatically miss the point. The point is rather that $\frac{1}{2}mv^2$ is a very small number relative to mc^2 because the velocities that we are used to dealing with are so small compared to the cosmic speed limit. The reason we live in our relatively low-energy existence is ultimately linked to the strengths of the forces of nature, particularly the relative weakness of the forces of electromagnetism and gravity. We will investigate this in more detail in Chapter 7, when we enter the world of particle physics.

It took humans around a half century after Einstein before they eventually figured out how to extract significant amounts of mass energy from matter, and the destruction of mass is exploited today by nuclear power plants. In stark contrast, nature has been exploiting $E = mc^2$ for billions of years. In a very real sense, it is the seed of life, for without it our sun would not burn and the earth would be shrouded forever in darkness.

6

And Why Should We Care?
Of Atoms, Mousetraps, and the Power of the Stars

We have seen how Einstein's famous equation forces us to reconsider the way we think about mass. We have come to appreciate that rather than being simply a measure of how much stuff something contains, mass is also a measure of the latent energy stored up within matter. We have also seen that if we could unlock it, then we would have a phenomenal source of energy at our disposal. In this chapter we will spend some time exploring the ways in which mass energy can actually be liberated. But before we turn to such useful practicalities, we would like to spend a little more time exploring our newfound equation, $E = mc^2 + \frac{1}{2}mv^2$, a little more carefully.

Remember, this version of $E = \gamma mc^2$ is only an approximation, although a pretty good one for speeds even as high as 20 percent of the speed of light. Writing it like this makes the separation into mass energy and kinetic energy most apparent, and we won't bother to remind you that it is just an approximation. Recall also that we can construct a vector in spacetime whose length in the space direction represents a conserved quantity, which reduces to the old-fashioned law of conservation of momentum for velocities that are small compared with the speed of light. Just as the length of the new spacetime momentum vector in the space direction is conserved, so too must its length in the time direction be a conserved quantity, and this length is $mc^2 + \frac{1}{2}mv^2$. We recognized that $\frac{1}{2}mv^2$ is the formula for a quantity long familiar to scientists, the kinetic energy, and so we identified the conserved quantity as energy. Very important, we didn't start off looking for the conservation of energy. It emerged quite unexpectedly when we were trying to find a spacetime version of the law of conservation of momentum.

Imagine a bucket of armed mousetraps, all storing energy in the springs. We know that wound-up springs store energy because when the trap is triggered there is a loud bang (which is energy being released as sound) and the trap might jump up in the air (energy being turned into kinetic energy). Now imagine that one trap goes off and triggers the rest. There is a huge clatter as the energy stored in the springs is liberated and the mousetraps snap shut. The conservation of energy says that the energy before the mousetraps snap shut must equal the energy afterward. Moreover, since the traps were initially all sitting at rest, the total energy must equal mc^2, where m is the total mass of the bucket of primed traps. Afterward, we have a bunch of spent traps *plus* the energy that was liberated. To balance the energy before with that afterward, it therefore follows that the bucket of armed mousetraps is actually more massive than the bucket of triggered traps. Let's think of another example, this time involving a contribution to mass arising from kinetic energy. A box full of hot gas has more mass than an identical box containing the same gas at a lower temperature. The temperature measures how fast the molecules are whizzing around inside the box—the hotter the gas, the faster the molecules move around. Because they are moving faster, they have more kinetic energy (i.e., the result of adding together the values of $\frac{1}{2}mv^2$ for each molecule is bigger for the hot gas) and hence the box has more mass. The logic extends to everything that stores energy. A new battery is more massive than a used battery, a hot flask of coffee is more massive than a cold one, and a steaming-hot meat and potato pie bought at halftime on a wet Saturday afternoon at Oldham Athletic's football ground is more massive than the same uneaten pie at the end of the game.

The conversion of mass to energy is therefore not such an exotic process. It is happening all the time. As you relax by a crackling fire you are absorbing heat from the burning coals, and that heat takes energy away from the coal. In the morning, when the fire has died away, you could very carefully sweep up every last piece of ash and weigh it with scales of unfeasible accuracy. Even if you miraculously managed to get every atom of ash, you would find that it weighed less than the original coals weighed. The difference would be equal to the amount of energy liberated divided by the speed of light squared, as predicted by $E = mc^2$, i.e., according to $m = E/c^2$. We can quickly figure out how tiny the change in mass would be for the kind of fire that might warm your house as the night draws near. If the fire generates 1,000 watts of power for 8 hours, then the total energy output is equal to 1,000 x (8 x 60 x 60) joules (because we have to work in seconds, not hours, in order to get an answer in joules), which is just less than 30 million joules. The corresponding loss of mass must therefore be equal to 30 million joules divided by the speed of light squared, and that is equal to less than one-millionth of a gram. The explanation for the tiny reduction in mass is a direct consequence of the conservation of energy. Before igniting the fire, the total energy of the coals is equal to the total mass of coal multiplied by the speed of light squared. As the fire burns, energy leaves the fire. Eventually, the fire dies and we are left with ash. According to the law of conservation of energy, the total energy of the ash must be less than the total energy of the coal by an amount equal to the energy that went into warming the room. The energy of the ash is equal to its mass multiplied by the speed of light squared, which must be lighter than the original coal by the amount we just calculated.

The process of converting mass into energy and energy into mass is therefore absolutely fundamental to the workings of nature; it really is an everyday occurrence. For anything to happen at all in the universe, energy and mass must be continually sloshing back and forth. How on earth did anyone manage to explain anything involving energy before we knew this seemingly most basic of facts about the workings of nature? It's worth remembering that Einstein first wrote down $E = mc^2$ in 1905 in a world that was far from primitive. The first intercity passenger railway, powered by coal-burning steam locomotives, was opened in 1830 between Liverpool and Manchester. Coal-burning ocean liners had been crossing the Atlantic for almost seventy years, and the golden age of steam was in full swing with advanced steam-turbine-powered liners, such as the *Mauretania* and *Titanic*, about to enter service. The Victorians certainly knew how to burn coal efficiently and to spectacular effect, but how did the scientists of the day think of the physics behind a burning fire before Einstein? A nineteenth-century engineer would have said the coal has latent energy stored within it (rather like the energy stored in lots of miniature mousetraps) and the chemical reactions that burn the coal spring the traps and liberate that energy. This picture works, and allows calculations to be made with the accuracy required to design a beautiful machine like an ocean liner or an express steam locomotive. The post-Einstein view does not disagree with this picture but rather it adds to it. That is to say, we now understand that latent energy is irrevocably intertwined with the concept of mass. The more latent energy something has, the more massive it is. It would not have occurred to scientists before Einstein that there was a link between mass and energy, because they had not been forced to think in that way. Their view of nature was accurate enough to explain the world they observed and to solve the problems they encountered, because the changes in mass were so tiny that they never needed to know them.

Here lies another insight into science. With each new level of understanding, a more accurate worldview emerges. The current worldview is never claimed to be correct, in the very important sense that there are no absolute truths in science. The body of scientific knowledge at any point in history, including now, is simply the collection of theories and views of the world that have not yet been shown to be wrong.

All of the examples we just looked at lead to very tiny fractional changes in mass, but of course the release of the corresponding energy can be very significant. A fire keeps us warm and a hot pie is much tastier than a cold one. In the case of burning coal, the stored energy is chemical in origin. The molecules that make up the coal get rearranged and turn to ash as a result of a chemical chain reaction initiated by a lighted match. As the bonds between the molecules snap and reform and atoms recombine with atoms to make new molecules, energy is released and the mass reduces. Chemical energy has its origins in the structure of atoms. The simplest example is a single hydrogen atom, which is a single electron in orbit around a single proton. It is simple enough that physicists can use the quantum theory to calculate how the mass of the atom should change as the electron moves around. There is a smallest value for the mass of a hydrogen atom. It is an utterly miniscule 0.00000000000000000000000000000000002 kilograms less than the combined mass of an electron and a proton that are far apart. Nevertheless, that difference, when converted into energy, is a very big deal. Ask any chemist or experience its effect yourself sitting in front of that nice coal fire.

Because particle physicists are as lazy as the next guy, they don't like writing very small numbers down with lots of zeros and decimal places, so they don't usually use kilograms to measure mass. Instead they use a unit called the electron volt, which is actually a measurement of energy. An electron volt is the amount of energy an electron gets when it is accelerated through a potential difference of 1 volt. This is a mouthful, and we are again in danger of covering ourselves in chalk dust. In more normal-sounding language, if you get a 9 volt battery and build a little particle accelerator out of it, you would be able to give an electron 9 electron volts of energy. The electron volt is turned into a mass by dividing it by c^2 (remember $E = mc^2$). In this rather more convenient language, the hydrogen atom has a smallest mass, which is 13.6 eV/c^2 less than the masses of the proton (938,272,013 eV/c^2) and electron (510,998 eV/c^2) combined (1 eV is the abbreviation for an energy of 1 electron volt). Notice that by keeping a factor of c^2 "in the units," it is easy to figure out how much energy is stored within a proton at rest. Since the energy is obtained by multiplying the mass by c^2, the c^2 factors cancel out and the energy is just 938,272,013 eV.

Notice also that the mass of a hydrogen atom is *smaller*, not bigger, than the sum of its component parts. It is as if the atom has some negative energy stored within it. There is nothing mystical about negative energy in this context: "Negative stored energy" just means that it takes effort to dismantle the atom, and it often goes by the name "binding energy." The next smallest mass of a hydrogen atom is 10.2 eV/c^2 smaller than the sum of its parts.* The mystical-sounding and oft-misunderstood quantum theory actually derives its name from the fact that masses like these come in discrete ("quantized") values. For example, there is no hydrogen atom with a mass 2 eV/c^2 bigger than the smallest mass. This is really all there is to the word "quantum." The different masses actually correspond to the electrons being in different orbits around the atomic nucleus, which in the case of hydrogen is a single proton.

That said, one has to be very careful in picturing electron orbits, because they are not really like the orbits of planets around the sun. Loosely speaking, the atom with the smallest mass has the electron closer to the proton than the atom with the next

---

* Strictly speaking, this is not true. There is another possible mass lying just 0.000006 eV/c^2 above the smallest mass. That tiny difference is very important to radio astronomers, but we will assume it is so close to the smallest mass that it makes no difference.

smallest mass, and so on. The hydrogen atom with the electron
as close as it can be to the proton is said to be in its "ground
state" and it is as light as it can be. Add just the right amount of
energy and the electron will jump up to the next available orbit
and the atom will become a bit heavier, simply because a bit of
energy has been added. In that sense, adding energy to an atom
is like winding up the spring in a mousetrap.

All of this does beg the question of how we know such fine
detail about hydrogen atoms. Surely we don't go around measuring these tiny mass differences using weighing scales? At
the heart of the quantum theory is an equation called the
Schrödinger wave equation, and we can use it to predict what
the masses should be. Legend has it that Schrödinger discovered the equation, one of the most important in modern
physics, while on a winter sojourn with his mistress in the Alps
over Christmas and New Year's of 1925–1926. Quite how he explained this to his wife is rarely discussed in physics textbooks.
We can only hope his mistress enjoyed the fruits of his labors as
much as the generations of physics students who know the
eponymous equation by heart. The calculation is not too difficult for an atom as simple as hydrogen, and it has graced many
an undergraduate examination paper. But mathematical
tractability means little without the corroborating evidence provided by experiments. Fortunately, the results of the quantum
nature of atomic structure are pretty easy to observe. In fact, we
all observe them every day. There is a general rule in quantum
theory that roughly goes like this: Left alone, a heavier thing will
turn into a lighter thing if at all possible. It is not a hard concept
to understand. If the thing is left alone it cannot possibly go to

a heavier thing because there is no energy being added, whereas there is always the chance it can shed some energy and become lighter. Of course, the third option is that it does nothing and stays the same, and sometimes that is the case. For the hydrogen atom this means that the heavier version will eventually shed some of its mass. It does so by emitting a single particle of light, the photon we met earlier. For example, a next-to-lightest hydrogen atom will at some point spontaneously convert into a lightest hydrogen atom as a consequence of a change in the orbit of the electron. The excess energy is carried away by a photon.* The reverse process can occur too. A photon, if one just happens to be around, can be absorbed by the atom, which then jumps to a higher mass because the energy absorbed promotes the electron to a higher orbit.

Perhaps the most everyday way of getting energy into atoms is to heat them up. This causes the electrons to jump up into the higher orbits and subsequently drop back down again, emitting photons as they go (this is the physics behind a sodium vapor street lamp). These photons carry an energy that is exactly equal to the energy difference between the orbits, and if we could detect them, we would have a direct window into the structure of matter. Fortunately, we are detecting them all the time because our eyes are nothing more (or less) than photon detectors, and the energy of the photons is registered directly as color. The azure blue of an island-pitted tropical ocean, the jagged diamond yellow of Van Gogh's stars, and the iron-red of your blood

---

* The energy taken away by the photon is equal to 13.6 eV minus 10.2 eV, which is 3.4 eV.

are a direct measurement by your eyes of the quantized structure of matter. The origin of the colors emitted by hot gases was one of the driving forces behind the discovery of quantum theory at the turn of the twentieth century. The years of careful observation of the light emitted from anything and everything by legions of diligent scientists are commemorated in our language by the name of the gas that fills party balloons. "Helium" is derived from the Greek word "helios," which means "sun," because the signature of this atom was first discovered by French astronomer Pierre Janssen in the light from a solar eclipse in 1868. In this way we discovered helium on our star before we found it on Earth. Today, astronomers search for signs of life on distant worlds by looking for the characteristic fingerprint of oxygen in the starlight shining through the atmospheres of planets as they pass across the face of their parent stars. Spectroscopy, as this branch of science is known, is a powerful tool for exploring the universe without and within.

All of the atoms in nature come in a tower of energies (or masses), depending on where the electrons are, and since there is more than a single electron in every atom except hydrogen, the light emitted from them spans all the colors of the rainbow and beyond, which is ultimately the reason why the world is so colorful. Chemistry is, very crudely, the area of science that is concerned with what happens when bunches of atoms come close together (but not too close). As two hydrogen atoms approach each other, the protons repel because they both carry positive electric charge, but that repulsion is overcome because the electron in one atom attracts the proton in the other. The result is that there is an optimal configuration where the two atoms are bound together to make a hydrogen molecule. The atoms are bound in the same sense that the electron is bound into orbit around a single hydrogen nucleus. Being bound means simply that it takes some effort to pull them apart and "it takes some effort" is a sloppy way of saying that we need to supply some energy. If we need to add energy just to break the molecule apart, then it follows that the molecule is less massive than the sum of the original two hydrogen atoms, just as the hydrogen atom is less massive than the sum of the masses of its constituents. In both cases, the binding energy comes about because of the force of electromagnetism that we met at the beginning of the book.

As everyone who has spent time in a school chemistry lab with a box of matches and an inattentive teacher knows, chemical reactions can sometimes lead to the production of energy. A coal fire is a perfect, nicely controlled example; a little nudge from a lighted match and energy is released steadily for hours. More dramatic, an exploding stick of dynamite releases similar amounts of energy to a coal fire, albeit rather more quickly. The energy doesn't come from the match that lit the fire or the fuse, but from the energy stored within. The bottom line is always that the combined mass of the products of the reaction must be less than the mass we started with if some energy has been lost.

A final example may serve to further illustrate the idea of energy release through chemical reactions. Imagine sitting in a room full of hydrogen and oxygen molecules. We would be able to breathe perfectly well, and at first sight it would appear quite safe and comfortable since it takes energy to pull apart two hydrogen atoms bound together in a molecule. This would seem to suggest that molecular hydrogen should be a stable substance. It can, however, be broken up via a chemical reaction that generates an impressive amount of energy; so impressive in fact that hydrogen gas is very dangerous stuff. It is highly flammable in air, needing only a tiny spark to trigger disaster. In our newfound language, we can analyze the process in a little more detail. Suppose we mix together a gas of hydrogen molecules (two hydrogen atoms bound together) and a gas of oxygen molecules (two oxygen atoms bound together). Now, you might well become very nervous sitting in your room when you discover that the combined mass of two hydrogen molecules and one oxygen molecule is *bigger* than the combined mass of two water molecules, each of which is made of two hydrogen atoms and an oxygen atom. In other words, the four hydrogen atoms and two oxygen atoms that started as molecules are more massive than two lots of H_2O. The excess mass is approximately $6 \ eV/c^2$. The hydrogen and oxygen molecules would therefore quite like to be rearranged into two water molecules. All that will be different is the configuration of the atoms (and their associated electrons). At first glance the energy release per molecule is tiny, but a roomful of gas contains in the region of 10^{26} molecules,* and that translates into around 10 million joules of energy, which is plenty enough to rearrange your own personal molecules as a side effect. Fortunately, if we are careful, then we are not destined to be incinerated because although the final products have a mass that is smaller than the initial products, it takes a bit of

---

* $10^1 = 10, 10^2 = 100$, etc. So 10^{26} is equal to 100000000000000000000000000 and you can see why the more compact notation was invented.

effort to put them, and their electrons, into the right configuration. It is a bit like pushing a bus over a cliff edge—it takes some effort to get it started but when it goes, there is no stopping it. That said, it would be very unwise to light a match, which would supply plenty enough energy to trigger the molecular rearrangement process and get the water production under way.

Liberating chemical energy by shuffling atoms around or gravitational energy by shuffling heavy things around (like huge volumes of water in hydroelectric plants) provides our civilization with a means to generate and harness energy. We are also becoming increasingly adept at harvesting the abundant resources of kinetic energy found in nature. As the wind blows, molecules of air rush along, and we can convert that wild kinetic energy into useful energy by putting a wind turbine in the way. The molecules bang into the blades of the turbine and as a result the molecules slow down, delivering their kinetic energy to the turbine, which starts to rotate (incidentally, that is another example of the conservation of momentum). In this way, the kinetic energy of the wind gets transformed into rotational energy of the turbine, and that in turn can be used to power a generator. Harnessing the power of the sea works in much the same way, except that it is the kinetic energy of water molecules that gets converted into useful energy. From a relativistic perspective, all forms of energy contribute to mass. Imagine a giant box filled with flying birds. You could put the box on a set of measuring scales and weigh it, thereby inferring the mass of the birds plus the box. Since the birds are flying around, they have some kinetic energy, and as a result the box will weigh a tiny bit more than it would if the birds were all asleep.

The energy released in chemical reactions has been the primary source of power for our civilization since prehistoric times. The amount of energy that can be liberated for a given amount of coal, oil, or hydrogen is at the most fundamental level determined by the strength of the electromagnetic force, since it is this force that determines the strength of the bonds between atoms and molecules that are broken and reformed in chemical reactions. However, there is another force of nature that offers the potential to deliver vastly more energy for a given amount of fuel, simply because it is much stronger.

Deep inside the atom lies the nucleus—a bunch of protons and neutrons stuck together by the glue of the strong nuclear force. Being glued together, it takes effort to pull a nucleus apart, just as it does for atoms and molecules, and its mass is therefore less than the sum of the masses of its individual proton and neutron parts. Entirely analogous to the goings-on in chemical reactions, we might wonder whether it is possible to make nuclei interact with each other in such a way that allows this mass difference to be emitted as useful energy. Breaking chemical bonds and releasing the stored energy in the atoms can be as easy to achieve as lighting a match, but releasing the energy bound up in a nucleus is an entirely different matter. It is often hard to access and usually requires some clever apparatus. Not always, though; there are occasions where nuclear energy is liberated naturally and spontaneously, with extremely important and unexpected consequences for planet Earth.

The heavy element uranium has 92 protons and, in its most stable naturally occurring form, 146 neutrons. In this guise, it has a half-life of around 4.5 billion years, which simply means that in 4.5 billion years, half of the atoms in a lump of uranium will have spontaneously split up into lighter things, the heaviest of these being the element lead, and liberated energy as a result. In the language of $E = mc^2$, the uranium nucleus splits into two smaller nuclei, whose combined mass is a little less than the mass of the original nucleus. It is that loss of mass that manifests itself as nuclear energy. The process whereby a heavy nucleus splits up into two lighter nuclei is called nuclear fission. Along with the 146-neutron form of uranium, there also exists a less-stable naturally occurring form with 143 neutrons that splits into a different form of lead with a half-life of 704 million years. These elements can be used to accurately date rocks almost as old as the earth itself, which is around 4.5 billion years old.

The technique is beautifully simple. There exists a mineral known as zircon that naturally incorporates uranium into its crystalline structure, but not lead. It can therefore be assumed that any lead present in the mineral comes from the radioactive decay of uranium, which allows the date of formation of the zircon to be measured with high precision simply by counting the number of lead nuclei present and knowing the rate of decay of the uranium. The heat generated when uranium splits up also plays a crucial role in keeping the earth warm, and that heat helps provide the power that drives plate tectonics and pushes up new mountains. Without this impetus, fueled by nuclear energy, the land would crumble into the sea as a result of natural erosion. We shall say no more about nuclear fission. It is now time to zoom in on the atomic nucleus and learn a little more about its stored energy and the other important process that can occur to facilitate its release: nuclear fusion.

Take two protons (no electrons are around this time, so we have no chance to make them stick together in a hydrogen molecule). Left alone, they would fly apart in opposite directions because they both carry positive electric charge. So it seems pretty pointless to try to push them closer together. Even so, let us imagine pushing the protons closer together and investigate what happens. One way to do this would be to hurl them at each other with increasing speed. The force of repulsion between the protons gets larger and larger as the protons get closer and closer together. In fact, it doubles in strength for every halving of the distance. It therefore seems that our protons are always destined to be flung apart. If the electrical repulsion were the only force in nature, this would certainly be the case. There are, however, the strong and weak nuclear forces to contend with. When the protons get so close together that they are almost touching each other (protons are not solid balls, so we can even think of them as overlapping) something very remarkable happens. Not always, but some of the time, when we bring two protons together like this, one of the protons will spontaneously turn itself into a neutron and the excess positive electric charge (the neutron being electrically neutral, hence its name) is shed as a particle called a positron. Positrons are identical to the electron except that they carry positive charge. Also emitted is a particle called a neutrino. Compared to the proton and neutron, which have very similar masses, the electron and neutrino are very light and they whiz off into the sunset, leaving the proton and neutron behind. The details of this transmutation process are very well understood using the theory of weak interactions developed by particle physicists in the second half of the twentieth century. We will show how it works in the next chapter. All we need to know here is that the process can and does occur. Free from the electric repulsion, the proton and neutron can snuggle together under the influence of the strong nuclear force. A proton and neutron bound up like that is called a deuteron, and the process of a proton turning into a neutron with the emission of a positron (or vice versa, with the emission of an electron, which can also happen) is called radioactive beta decay.

How does all of that fit with our understanding of energy? Well, the two original protons each have a mass of 938.3 MeV/c^2. 1 MeV is equal to 1 million eV (the "M" stands for "mega" or "million"). The conversion between MeV/c^2 and kilograms is easy enough: 938.3 MeV/c^2 corresponds to a mass of 1.673 x 10^{-27} kilograms.* The two original protons have a total mass of 1876.6 MeV/c^2. The deuteron has a mass of 1875.6 MeV/c^2, and the energy associated with the 1 McV remainder is carried away by the positron and neutrino, of which approximately half is used up to manufacture the positron since it has a mass of around ½ MeV/c^2 (neutrinos have almost no mass at all). So when two protons convert into a deuteron, a relatively tiny fraction (around 1/40 of 1 percent) of the total mass is destroyed and converted into the kinetic energy of the positron and the neutrino.

Squeezing two protons together to make a deuteron is one way to liberate the energy bound up in the strong force, and it is

---

* 10^{-1} = 0.1, 10^{-2} = 0.01, etc. So 10^{-27} has twenty-six zeros after the decimal point.

an example of nuclear fusion. The term "fusion" is used to describe any process that releases energy as a result of fusing together two or more nuclei. In contrast to the energy released in a chemical reaction, which is a result of the electromagnetic force, the strong nuclear force generates a huge binding energy. For example, compare the ½ MeV released when a deuteron is formed to the 6 eV released in our hydrogen-oxygen explosion. This is in keeping: The energy released in a nuclear reaction is typically a million times the energy released in a chemical reaction. The reason that fusion doesn't happen all the time in our everyday experience here on Earth is that, because the strong force operates only over short distances, it only kicks in when the constituents are very close together and declines very rapidly at distances much greater than a femtometer (which is roughly equal to the size of one proton). But it is not easy to push protons together to that distance because of their electromagnetic repulsion. One way to do it requires the protons to be moving extremely fast, and this in turn means a very high temperature indeed because temperature is essentially nothing more than a measure of the average speed of things; the molecules of water in a hot cup of tea are jiggling around more than the molecules in a cold pint of beer. At the very least a temperature of around 10 million degrees is necessary for fusion to begin, and preferably significantly more. Fortunately for us, there are places in the universe where temperatures meet and exceed those necessary for nuclear fusion—deep in the hearts of stars.

Let us journey back in time to the cosmic dark age, less than half a billion years after the big bang when the universe is filled with only hydrogen, helium, and a sprinkling of the lighter chemical elements. Slowly, as the universe continues to expand and cool, the primordial gases begin to fall in on themselves in clumps under the influence of gravity, picking up speed as they rush toward each other, just as this book will speed up toward the ground if you drop it. Faster-moving hydrogen and helium means hotter hydrogen and helium, so the big balls of gas become increasingly hot and increasingly dense. At a temperature of 10,000 degrees, the electrons are ripped from their orbits around the nuclei, leaving behind a gas of protons and electrons known as a plasma. Together the individual electrons and protons continue to fall inexorably inward, faster and faster in a relentlessly quickening collapse. The plasma is rescued from a seemingly irretrievable fall when the temperature approaches 10 million degrees, when something very important happens, something that transforms the hot ball of protons and electrons into the life and light of the universe; a magnificent source of nuclear energy; a star. Individual protons fuse together to make a deuteron, which itself can fuse with another proton to produce helium, and all the while precious binding energy is released. In this way the new star slowly converts a small fraction of the original mass into energy, which heats up the core of the star and allows it to halt and resist any further gravitational collapse, at least for a few billion years—time enough for cold, rocky planets to be warmed, liquid water to flow, animals to evolve, and civilizations to rise.

Our sun is a star that is currently in just such a comfortable midlife phase: It is burning hydrogen to make helium. In the process, it loses 4 million tons of mass every second of every day of every millennium as it converts 600 million tons per second of hydrogen into helium. This profligacy, the foundation of our existence, cannot continue forever, even for our local ball of plasma, large enough to contain a million earths. So what happens when a star runs out of hydrogen fuel in its core? Without the nuclear source of outward pressure, the star will once again start to collapse, getting hotter and hotter as it does so. Eventually, at a temperature of around 100 million degrees, helium begins to burn and once again the star's collapse is arrested. We are using the word "burn," but that isn't really very precise. What we really mean is that nuclear fusion is taking place and the net mass of the final products is less than the mass of the original fusing material—the loss of mass leading to the production of energy in accord with $E = mc^2$.

The process of burning helium is really worth a closer look. When two helium nuclei fuse, they make a particular form of beryllium, made up of four protons and four neutrons. This form, called beryllium-8, lives for only one ten-millionth-of-a-billionth of a second before it falls apart into two helium nuclei again. The brief life of beryllium-8 is so fleeting that it is very unlikely it will hang around long enough to fuse with anything else. In fact, without a helping hand, that is pretty much what would always happen, and the pathway to synthesizing heavier elements inside stars would be blocked. In 1953, when the understanding of the nuclear physics of stars was still in its infancy, astronomer Fred Hoyle realized that carbon had to be manufactured inside stars, irrespective of what the nuclear physicists told him, because he strongly believed that there is nowhere else in the universe to make it. Coupled with his astute observation that astronomers exist, he theorized that this could happen only if a slightly heavier type of carbon nucleus exists such that it can be formed very efficiently as the result of fusion between the short-lived beryllium-8 and a third helium nucleus. For the theory to work out, Hoyle figured out that the heavy carbon should be 7.7 MeV/c^2 heavier than ordinary carbon. Once this new form of carbon has been made in the star, the pathway to heavier elements opens up. At the time, no such form of carbon was known but, spurred on by Hoyle's prediction, scientists wasted no time in hunting for it. It was a matter of days after Hoyle made his prediction that nuclear physicists working in the Kellogg Laboratory at Caltech confirmed his prediction without any shadow of doubt. This is a remarkable story, not least because of the way it helps us build confidence in our understanding of how stars work: There is no better vindication of a beautiful theory than the verification in an experiment of a prior prediction.

Today we have a great deal more evidence that supports the theory of stellar evolution. One striking example comes from the study of the neutrinos produced every time a proton turns into a neutron in the fusion process. Neutrinos are ghostly particles that hardly ever interact with anything, and as such, most of them stream out from the sun as soon as they are produced without hindrance. The neutrino flux is so great, in fact, that around 100 billion of them pass through each square centimeter of the earth every second. This is an easy fact to read but an astonishing thing to imagine. Hold your hand up in front of you and look at your thumbnail. Each second, 100 billion subatomic

particles from the core of our star will pass through it. Fortunately for us, the neutrinos nearly always pass through our
hands, and in fact the entire earth, as if they did not exist. However, on rare occasions, a neutrino will interact, and the trick is
to build experiments that are able to catch these extremely rare
events. The Super-Kamiokande experiment, located deep in the
Mozumi mine near the city of Hida in Japan, is up to the challenge. Super-Kamiokande is a huge cylinder 40 meters across
and 40 meters tall, containing 50,000 tons of pure water, surrounded by over 10,000 photomultiplier tubes that are capable
of detecting the very faint flashes of light that are produced
when a neutrino collides with an electron in the water. As a result, the experiment is able to "see" the neutrinos streaming
from the sun, and the number arriving turns out to agree with
expectations based upon the theory that they are produced as a
result of fusion processes inside the sun.

Eventually, the star will exhaust its supply of helium and
begin to collapse even further. As the core temperature rises past
500 million degrees, it becomes possible for the carbon to burn,
producing a variety of heavier elements all the way up to iron.
Your blood is red because it contains iron, the end point of fusion in the core of stars. Elements heavier than iron cannot be
manufactured through fusion in the core because there is a law
of diminishing returns, and for nuclei heavier than iron there is
no more energy to be released from fusing with extra nuclei. In
other words, adding protons or neutrons to an iron nucleus can
only make it heavier (not lighter, as would be necessary for fusion to act as a source of energy). Nuclei heavier than iron pre-

fer instead to shed protons or neutrons, as we saw earlier in the case of uranium. In these cases, the sum total of the masses of the products is less than the mass of the initial nucleus, and so energy is released when a heavy nucleus divides. Iron is the special case; it is the Goldilocks nucleus and that means that iron is exceptionally stable.

With no other source of energy available to prevent the inevitable, a star that has an iron-rich core is really at the point of no return, and gravity resumes its relentless work. There is now only one last chance for the star to prevent total collapse. It becomes so dense that the electrons that have been hanging around ever since they were ripped off the hydrogen atoms during its birth resist further collapse as a result of the Pauli exclusion principle. The principle is an important one in quantum theory and it is crucial for the stability and structure of atoms. Crudely put, it says that there is a limit to how closely you can pack electrons together. In a dense star, the electrons exert an outward pressure that increases as the star collapses until it is eventually so large that it can prevent any further gravitational collapse. Once that happens, the star is trapped in an enfeebled but incredibly long-lived state. It has no fuel to burn (that is why it was collapsing in the first place) and it cannot collapse any further because of the electron pressure. Such a star is called a white dwarf—a slowly fading memorial to a majesty irredeemably diminished—the once-bright creator of the elements of life compressed into a remnant the size of a small planet. In a time far longer than the age of the universe today, the white dwarfs will have cooled so much that they fade from view. We are reminded of the beautiful sentiments of the father of the big bang theory, Georges Lemaitre, when reflecting on the inevitable universal journey from light into darkness from which even stars cannot escape: "The evolution of the universe can be likened to a display of fireworks that has just ended: some few wisps, ashes and smoke. Standing on a well-cooled cinder, we see the fading of the suns, and try to recall the vanished brilliance of the origins of the worlds."

It has been our goal throughout this book to be careful to explain why things are as they are and to provide arguments and evidence as we progress. The description we presented here of how a star works might seem fanciful, and we have certainly deviated from our careful, explanatory style. You might even object that since it is not possible to do laboratory experiments directly on stars, we cannot possibly be certain how they work. But that isn't why we were brief. We have been brief because it would take us too far from the point to go into more detail. The remarkable work of Hoyle and the success of experiments like Super-Kamiokande will have to suffice by way of supporting evidence, along with one last beautiful prediction made by Indian physicist Subrahmanyan Chandrasekhar. In the early 1930s, armed only with already well-established physics, he predicted that there should be a largest possible mass for any (nonrotating) white dwarf star. Chandrasekhar originally estimated the largest mass to be around 1 solar mass (i.e., the mass of the sun), and more refined calculations later led to a value of 1.4 solar masses. At the time of Chandrasekhar's work, only a handful of white dwarf stars had been observed. Today, around 10,000 white dwarf stars have been observed, and they typically have a mass close to that of the sun. Not a single one has a mass that exceeds Chandrasekhar's maximum value. It is one of the true joys of physics that laws discovered in tabletop experiments in a darkened laboratory on earth pertain throughout the universe, and Chandrasekhar exploited that universality to make his prediction. For that work he received the 1983 Nobel Prize. The validation of his prediction is one of the pieces of evidence that allows physicists to be very confident that they really know how stars work.

Are all stars fated to end their lives as white dwarf stars? The narrative in the previous paragraph suggests so, but it is not the whole story and there was a clue. If there can never be a white dwarf star with a mass larger than 1.4 solar masses, what happens to stars that are bigger than that? Putting aside the possibility that big stars can shed material so that they sneak in under Chandrasekhar's limit, two alternative fates await. In both cases, the large initial mass means that the electrons eventually start to move around at close to the speed of light as the collapse continues. Once that happens, there really is nowhere else to go; their pressure will never be sufficient to resist the force of gravity. For these massive stars, the next stop is a neutron star, in which nuclear fusion steps in for a final time. The protons and electrons move so fast that they reach a point where they have sufficient energy to initiate proton-electron fusion, producing a neutron. The reaction is the reverse of the radioactive beta decay process, whereby a neutron spontaneously decays into a proton and an electron with the emission of a neutrino. In this

way, all of the protons and electrons gradually convert into neutrons and the star is nothing but a ball of neutrons. The density
of a neutron star is phenomenal: A single teaspoon of neutron
star matter weighs more than a mountain. Neutron stars are
stars that are more massive than our sun yet are compressed to
the size of a city.* Many of the known neutron stars spin at phenomenal rates and blast beams of radiation out into space like
cosmic lighthouses. These stars are known as pulsars, and they
are truly wonders of the universe. Some known pulsars are approaching twice the mass of our sun, measure only 20 kilometers in diameter, and spin more than five hundred times every
second. Imagine the violence of the forces on such an object.
We have discovered wonders beyond imagination.

Beyond neutron stars, a final fate awaits the biggest stars. Just
as the electrons can approach the speed of light in white dwarfs,
the neutrons in a neutron star can bump up against the limit
Einstein imposed on them. When this happens, no known force
will prevent complete collapse, and the star is destined to form
a black hole. Today our knowledge of the physics of space and
time inside black holes is incomplete. As we shall see in the final
chapter, the presence of mass causes spacetime to warp away
from the Minkowski spacetime that we have become so familiar with, and in the case of a black hole, that warping is so extreme that not even light can escape its clutches. In such extreme
environments, the laws of physics as we currently know them

---

* The largest mass of a neutron star can be estimated in a manner similar
to Chandrasekhar's limit for the largest possible mass of a white dwarf—i.e., by
assuming that the neutrons do not travel close to the speed of light if they are
to form a neutron star.

break down, and figuring out the way forward is one of the great challenges for twenty first-century science, for only then will we be able to complete the story of the stars.

7

The Origin of Mass

The discovery of $E = mc^2$ marked a turning point in the way physicists viewed energy, for it taught us to appreciate that there is a vast latent energy store locked away inside mass itself. It is a store of energy much greater than anyone had previously dared imagine: The energy locked away in the mass of a single proton is approaching 1 billion times what is liberated in a typical chemical reaction. At first sight it seems we have the solution to the world's energy problems, and to a degree that may well be the case in the long term. But there is a fly in the ointment, and a big one too: It is very hard to destroy mass completely. In the case of a nuclear fission power plant, only a very tiny fraction of the original fuel is actually destroyed; the rest is converted into lighter elements, some of which may be highly toxic waste products. Even within the sun, fusion processes are remarkably ineffective at converting mass into energy, and this is not only because the fraction of mass that is destroyed is very small: For any particular proton, the chances of fusion ever taking place are exceedingly remote because the initial step of converting a proton into a neutron is an incredibly rare occurrence—so rare, in fact, that it takes around 5 billion years on average before a proton in the core of the sun fuses with another proton to make a deuteron, thereby triggering the release of energy. Actually, the process would never even occur if it weren't for the fact that the quantum theory reigns supreme at such small distances: In the pre-quantum worldview, the sun is simply not hot enough to push the protons close enough together for fusion to take place—it would have to be around 1,000 times hotter than its current core temperature of 10 million degrees. When the British physicist Sir Arthur Eddington first proposed that fusion might be the power source of the sun in 1920, he was quickly made aware of this potential problem with his theory. Eddington was quite sure that hydrogen fusion into helium was the power source, however, and that an answer to the conundrum of the low temperature would soon be found. "The helium which we handle must have been put together at some time and some place," he said. "We do not argue with the critic who urges that the stars are not hot enough for this process; we tell him to go and find a hotter place."

So ponderous is the conversion of protons into neutrons that, "kilogram for kilogram," the sun is several thousand times less efficient than the human body at converting mass to energy. One kilogram of the sun generates only 1/5,000 of a watt of power on average, whereas the human body typically generates somewhat more than 1 watt per kilogram. The sun is of course very big, which more than makes up for its relative inefficiency.

As we have been so keen to emphasize in this book, nature works according to laws. So it will not do to get too excited about an equation that tells us, as $E = mc^2$ does, about what *might possibly* happen. There is a world of a difference between our imagination and what actually happens, and although $E = mc^2$ excites us with its possibilities, we must still understand just how it is that the laws of physics allow mass to be destroyed and energy released. Certainly the equation itself does not logically imply that we have a right to convert mass to energy at will.

One of the wonderful developments in physics over the past hundred years or so has been the realization that we appear to need only a handful of laws to explain pretty much all of physics—at least in principle. Newton seemed to have achieved that goal when he wrote down his laws of motion way back in the late seventeenth century, and for the next two hundred years there was little scientific evidence to the contrary. On that matter, Newton was rather more modest. He once said, "I was like a boy playing on the sea-shore, and diverting myself now and then finding a smoother pebble or a prettier shell than ordinary, whilst the great ocean of truth lay all undiscovered before me," which beautifully captures the modest wonder that time spent doing physics can generate. Faced with the beauty of nature, it seems hardly necessary, not to mention foolhardy, to lay claim to having found the ultimate theory. Notwithstanding this appropriate philosophical modesty about the scientific enterprise, the post-Newton worldview held that everything might be made up of little parts that dutifully obeyed the laws of physics as articulated by Newton. There were admittedly some apparently minor unanswered questions: How do things *actually* stick together? What are the tiny little parts *actually* made of? But few people doubted that Newton's theory sat at the heart of everything—the rest was presumed to be a matter of filling in the details. As the nineteenth century progressed, however, there came to be observed new phenomena whose description defied Newton and eventually opened the doors to Einstein's relativity and the quantum theory. Newton was duly overturned or, more accurately, shown to be an approximation to a more accurate view of nature, and one hundred years later we sit here again, perhaps ignoring the lessons of the past and claiming that we (almost) have a theory of all natural phenomena. We may well be wrong again, and that would be no bad thing. It is worth remembering not only that scientific hubris has often been shown to be folly in the past, but also that the perception that we somehow know enough, or even all there is to know, about the workings of nature has been and will probably always be damaging to the human spirit. In a public lecture in 1810, Humphry Davy put it beautifully: "Nothing is so fatal to the progress of the human mind as to suppose our views of science are ultimate; that there are no new mysteries in nature; that our triumphs are complete; and that there are no new worlds to conquer."

Perhaps the whole of physics as we know it represents only the tip of the iceberg, or maybe we really are closing in on a "theory of everything." Whichever is the case, one thing is certain: We currently have a theory that is demonstrably proven, after a vast and painstaking effort by thousands of scientists around the world, to work across a very broad range of phenomena. It is an astonishing theory, for it unifies so much, yet its central equation can be written on the back of an envelope.

$$L = -\frac{1}{4}W_{\mu\nu}W^{\mu\nu} - \frac{1}{4}B_{\mu\nu}B^{\mu\nu} - \frac{1}{4}G_{\mu\nu}G^{\mu\nu}$$
$$+\overline{\psi}_j\gamma^\mu\left(i\partial_\mu - g\boldsymbol{\tau}_j\cdot\mathbf{W}_\mu - g'Y_jB_\mu - g_s\mathbf{T}_j\cdot\mathbf{G}_\mu\right)\psi_j$$
$$+|D_\mu\phi|^2 + \mu^2|\phi|^2 - \lambda|\phi|^4$$
$$-\left(y_j\overline{\psi}_{jL}\phi\,\psi_{jR} + y'_j\overline{\psi}_{jL}\phi_c\,\psi_{jR} + \text{ conjugate}\right)$$

FORMULA 7.1

We'll call this central equation the master equation, and it lies at the heart of what is now known as the Standard Model of Particle Physics. Although it is unlikely to mean much to most readers at first sight, we can't resist showing it above.

Of course, only professional physicists are going to know what's going on in detail in the equation, but we did not show it for them. First, we wanted to show one of the most wonderful equations in physics—in a moment we will spend quite some time explaining why it is so wonderful. But also it really is possible to get a flavor of what is going on just by talking about the symbols without knowing any mathematics at all. Let us warm up by first describing the scope of the master equation: What is its job? What does it do? Its job is to specify the rules according to which every particle in the entire universe interacts with every other particle. The sole exception is that it does not account for gravity, and that is much to everyone's chagrin. Gravity notwithstanding, its scope is still admirably ambitious. Figuring out the master equation is without doubt one of the great achievements in the history of physics.

Let's be clear what we mean when two particles interact. We mean that something happens to the motion of the particles as a result of their interaction with each other. For example, two particles could scatter off each other, changing direction as they do so. Or perhaps they might spin into orbit around each other, each trapping the other into what physicists call a "bound state." An atom is an example of such a thing, and in the case of hydrogen, a single electron and a single proton are bound together according to the rules laid down in the master equation. We heard a lot about binding energy earlier in the previous chapter, and the rules for how to calculate the binding energy of an atom, molecule, or atomic nucleus are contained in the master equation. In a sense, knowing the rules of the game means we are describing the way the universe operates at a very fundamental level. So what are the particles out of which everything is made, and just how do they interact with each other?

The Standard Model takes as its starting point the existence of matter. More precisely, it assumes the existence of six types of "quark," three types of "charged lepton," of which the electron is one, and three types of "neutrino." You can see the matter particles as they appear in the master equation: They are denoted by the symbol ψ (pronounced "psi"). For every particle there should also exist a corresponding antiparticle. Antimatter is not the stuff of science fiction; it is a necessary ingredient of the universe. It was British theoretical physicist Paul Dirac who first realized the need for antimatter in the late 1920s when he predicted the existence of a partner to the electron called the positron, which should have exactly the same mass but opposite electrical charge. We have met positrons before as the byproducts of the process whereby two protons fuse to make the deuteron. One of the wonderfully convincing features of a successful scientific theory is its ability to predict something that has never before been seen. The subsequent observation of that "something" in an experiment provides compelling evidence that we have understood something real about the workings of the universe. Taking the point a little further, the more predictions a theory can make, then the more impressed we should be if future experiments vindicate the theory. Conversely, if experiments do not find the thing that is predicted, then the theory cannot be right and it needs to be ditched. There is no room for debate in this kind of intellectual pursuit: Experiment is the final arbiter. Dirac's moment of glory came just a few years later when Carl Anderson made the first direct observations of positrons using cosmic rays. For their efforts, Dirac shared the 1933 Nobel Prize and Anderson the 1936 prize. Esoteric though the positron might appear to be, its existence is today used routinely in hospitals all over the world. PET scanners (short for "positron emission tomography") exploit positrons to allow doctors to construct three-dimensional maps of the body. It is not likely that Dirac had medical imaging applications in mind when he was wrestling with the idea of antimatter. Once again it seems that understanding the inner workings of the universe turns out to be useful.

There is one other particle that is presumed to exist, but it would be to rush things to mention it just yet. It is represented by the Greek symbol ϕ (pronounced "phi") and it is lurking on the third and fourth lines of the master equation. Apart from this "other particle," all of the quarks, charged leptons, and neutrinos (and their antimatter partners) have been seen in experiments. Not with human eyes, of course, but most recently with particle detectors, akin to high-resolution cameras that can take a snapshot of the elementary particles as they fleetingly come into existence. Very often, spotting one of them has won a Nobel Prize. The last to be discovered was the tau neutrino in the year 2000. This ghostly cousin of the electron neutrinos that stream out of the sun as a result of the fusion process completed the twelve known particles of matter.

The lightest of the quarks are called "up" and "down," and protons and neutrons are built out of them. Protons are made mainly of two up quarks and one down, while neutrons are made from two downs and one up. Everyday matter is made of atoms, and atoms consist of a nuclear core, made from protons and neutrons, surrounded at a relatively large distance by some electrons. As a result, up and down quarks, along with the electrons, are the predominant particles in everyday matter. By the way, the names of the particles have absolutely no technical significance at all. The word "quark" was taken from *Finnegan's Wake*, a novel by Irish novelist James Joyce, by American physicist Murray Gell-Mann. Gell-Mann needed three quarks to explain the then known particles, and a little passage from Joyce seemed appropriate:

> Three quarks for Muster Mark!
> Sure he has not got much of a bark
> And sure any he has it's all beside the mark.

Gell-Mann has since written that he originally intended the word to be pronounced "qwork," and in fact had the sound in his mind before he came across the *Finnegan's Wake* quotation.

Since "quark" in this rhyme is clearly intended to rhyme with "Mark" and "bark," this proved somewhat problematic. Gell-Mann therefore decided to argue that the word may mean "quart," as in a measure of drink, rather than the more usual "cry of a gull," thereby allowing him to keep his original pronunciation. Perhaps we will never really know how to pronounce it. The discovery of three more quarks, culminating in the top quark in 1995, has served to render the etymology even more inappropriate, and perhaps should serve as a lesson for future physicists who wish to seek obscure literary references to name their discoveries.

Despite his naming tribulations, Gell-Mann was proved correct in his hypothesis that protons and neutrons are built of smaller objects, when the quarks were finally glimpsed at a particle accelerator in Stanford, California, in 1968, four years after the original theoretical prediction. Both Gell-Mann and the experimenters who uncovered the evidence were subsequently awarded the Nobel Prize for their efforts.

Apart from the matter particles that we have just been talking about, and the mysterious ϕ, there are some other particles we need to mention. They are the W and Z particles, the photon and the gluon. We should say an introductory word or two about their role in affairs. These are the particles that are responsible for the interactions between all the other particles. If they did not exist, then nothing in the universe would ever interact with anything else. Such a universe would therefore be an astonishingly dull place. We say that their job is to carry the force of interaction between the matter particles. The photon is the particle responsible for carrying the force between electrically charged particles like the electrons and quarks. In a very real sense it underpins all of the physics uncovered by Faraday and Maxwell and, as a bonus, it makes up visible light, radio waves, infrared and microwaves, X-rays, and gamma rays. It is perfectly correct to imagine a stream of photons being emitted by a lightbulb, bouncing off the page of this book and streaming into your eyes, which are nothing more than sophisticated photon detectors. A physicist would say that the photon mediates the electromagnetic force. The gluon is not as pervasive in everyday life as the ubiquitous photon, but its role is no less important. At the core of every atom lies the atomic nucleus. The nucleus is a ball of positive electric charge (recall that the protons are all electrically charged, while the neutrons are not) and, in a manner analogous to what happens when you try to push two like poles of a magnet together, the protons all repel each other as a result of the electromagnetic force. They simply do not want to stick together and would much rather fly apart. Fortunately, this does not happen, and atoms exist. The gluon mediates the force that "glues" together the protons inside the nucleus, hence the silly name. The gluon is also responsible for holding the quarks together inside the protons and neutrons. This force has to be strong enough to overcome the electromagnetic force of repulsion between the protons, and for that reason it is called the strong force. We are really not covering ourselves in glory in the naming-stakes.

The W and Z particles can be bundled together for our purposes. Without them the stars would not shine. The W particle in particular is responsible for the interaction that turns a proton into a neutron during the formation of the deuteron in the core of our sun. Turning protons into neutrons (and vice versa) is not the only thing the weak force does. It is responsible for hundreds of different interactions among the elementary particles of nature, many of which have been studied in such experiments as those carried out at CERN. Apart from the fact that the sun shines, the *W* and *Z* are rather like the gluon in that they are not so apparent in everyday life. The neutrinos only ever interact via the *W* and *Z* particles and because of that they are very elusive indeed. As we saw in the last chapter, many billions of them are streaming through your head every second, and you don't feel a thing because the force carried by the *W* and *Z* particles is extremely weak. You've probably already guessed that we've named it the weak force.

So far we have done little more than trot off a list of which particles "live" in the master equation. The twelve matter particles must be added into the theory a priori, and we don't really know why there are twelve of them. We do have evidence from observations of the way that *Z* particles decay into neutrinos made at CERN in the 1990s that there are no more than twelve, but since it seems necessary to have only four (the up and down quarks, the electron, and the electron neutrino) to build a universe, the existence of the other eight is a bit of a mystery. We suspect that they played an important role in the very early universe, but exactly how they have been or are involved in our existence today is something to be added to the big unanswered questions in physics. Humphry Davy can rest easy for the moment.

As far as the Standard Model goes, the twelve are all *elementary* particles, by which we mean that the particles cannot be split up into smaller parts; they are the ultimate building blocks. That does seem to go against the grain of common sense—it seems perfectly natural to suppose that a little particle could, in principle, be chopped in half. But quantum theory doesn't work like that—once again our common sense is not a good guide to fundamental physics. As far as the Standard Model goes, the particles have no substructure. They are said to be "pointlike" and that is the end of the matter. In due course, it might well turn out that an experiment reveals that quarks can be split into smaller parts, but the point is that it does not have to be like that; pointlike particles could be the end of the story and questions of substructure might be meaningless. In short, we have a whole bunch of particles that make up our world and the master equation is the key to understanding how they all interact with each other.

One subtlety we haven't mentioned is that although we keep speaking of particles, it really is something of a misnomer. These are not particles in the usual sense of the word. They don't go around bouncing off each other like miniature billiard balls. Instead they interact with each other much more like the way surface waves can interact to produce shadows on the bottom of a swimming pool. It is as if the particles have a wavelike character while remaining particles nonetheless. This is again a very counterintuitive picture and it arises out of the quantum theory. It is the precise nature of those wavelike interactions that is rigorously (i.e., mathematically) specified by the master equation. But how did we know what to write down when we wrote the master equation? According to what principles does it arise? Before tackling these obviously very important questions, let's look a little more deeply at the master equation and try to gain some insight into what it actually means.

The first line represents the kinetic energy carried by the W and Z particles, the photon and the gluon, and it tells us how they interact with each other. We didn't mention that possibility yet but it is there: Gluons can interact with other gluons and W and Z particles can interact with each other; the W can also interact with the photon. Missing from the list is the possibility that photons can interact with photons, because they do not interact with each other. It is fortunate that they don't, because if they did it would be very difficult to see things. In a sense it is a remarkable fact that you can read this book. The remarkable thing is that the light coming from the page does not get bounced off-track on the way to your eyes by all the light that cuts across it from all the other things around you, things you could see if you turned your head. The photons literally slip past, oblivious to each other.

The second line of the master equation is where much of the action is. It tells us how every matter particle in the universe interacts with every other one. It contains the interactions that are mediated by the photons, the W and Z particles, and the gluons. The second line also contains the kinetic energies of all the matter particles. We'll leave the third and fourth lines for the time being.

As we have stressed, buried within the master equation are, bar gravity, all the fundamental laws of physics we know of. The law of electrostatic repulsion, as quantified by Charles Augustin de Coulomb in the late eighteenth century is in there (lurking in the first two lines), as is the entirety of electricity and magnetism, for that matter. All of Faraday's understanding and Maxwell's beautiful equations just appear when we "ask" the master equation how the particles with electric charge interact with each other. And of course, the whole structure rests firmly on Einstein's special theory of relativity. In fact, the part of the Standard Model that explains how light and matter interact is called quantum electrodynamics. The "quantum" reminds us that Maxwell's equations had to be modified by the quantum theory. The modifications are usually very tiny and lead to sub-tle effects that were first explored in the middle of the twentieth century by Richard Feynman and others. As we have seen, the master equation also contains the physics of the strong and weak forces. The properties of these three forces of nature are specified in all of their details, which means that the rules of the game are laid out with mathematical precision and without am-biguity or redundancy. So, apart from gravity, we seem to have something approaching a grand unified theory. It is certainly the case that no one has ever found any evidence anywhere in any experiment or through any observation of the cosmos that there is a fifth force at work in the universe. Most everyday phe-nomena can be explained pretty thoroughly using the laws of electromagnetism and gravity. The weak force keeps the sun burning but otherwise is not much experienced on Earth in everyday life, and the strong force keeps atomic nuclei intact but extends barely outside of the nucleus, so its immense strength does not reach out into our macroscopic world. The il-lusion that such solid things as tables and chairs are actually solid is provided by the electromagnetic force. In reality, matter is mainly empty space. Imagine zooming in on an atom so that the nucleus is the size of a pea. The electrons might be grains of sand whizzing around at high speeds a kilometer or so away—the rest is emptiness. The "grain of sand" analogy is stretching the point a little, for we should remember that they act rather more like waves than grains of sand, but the point here is to emphasize the relative size of the atom compared to the size of the nucleus at its core. Solidity arises when we try to push the cloud of electrons whizzing around the nucleus through the cloud of a neighboring atom. Since the electrons are electrically charged, the clouds repel and prevent the atoms from passing through each other, even though they are largely empty space. A big clue to the emptiness of matter comes when we look through a glass window. Although it feels solid, light has no trouble passing through, allowing us to see the outside world. In a sense, the real surprise is why a block of wood is opaque rather than transparent!

It is certainly impressive that we can shoehorn so much physics into one equation. It speaks volumes for Wigner's "unreasonable effectiveness of mathematics." Why should the natural world not be far more complex? Why do we have a right to condense so much physics into one equation like that? Why should we not need to catalog everything in huge databases and encyclopedias? Nobody really knows why nature allows itself to be summarized in this way, and it is certainly true that this apparent underlying elegance and simplicity is one of the reasons why many physicists do what they do. While reminding ourselves that nature may not continue to submit itself to this wonderful simplification, we can at least for the moment marvel at the underlying beauty we have discovered.

Having said all that, we are still not done. We haven't yet mentioned the crowning glory of the Standard Model. Not only does it include within it the electromagnetic, strong, and weak interactions, but it also unifies two of them. Electromagnetic phenomena and weak interaction phenomena at first sight appear to have nothing to do with each other. Electromagnetism is the archetypal real-world phenomenon for which we all have an intuitive feel, and the weak force remains buried in a murky subnuclear world. Yet remarkably the Standard Model tells us that they are in fact different manifestations of the same thing. Look again at the second line of the master equation. Without knowing any mathematics, you can "see" the interactions between matter particles. The portions of the second line involving W, B, and G (for gluon) are sandwiched between two matter particles, ψ, and that means that here are the bits of the master equation that tell us how matter particles "couple" with the force mediators but with a punch line. The photon lives partly in the symbol "W" and partly in "B," and that is where the Z lives too! The W particle lives entirely in "W." It is as if the mathematics regards the fundamental objects as W and B, but they mix up to conjure the photon and the Z. The result is that the electromagnetic force (mediated by the photon) and the weak force (mediated by the W and Z particles) are intertwined. In experiments, it means that properties that can be measured in experiments on electromagnetic phenomena should be related to properties measured in experiments on weak phenomena. That is a very impressive prediction of the Standard Model. And it was a prediction: The architects of the Standard Model, Sheldon Glashow, Steven Weinberg, and Abdus Salam, shared a

Nobel Prize for their efforts, for their theory was able to predict the masses of the W and Z particles well before they were discovered at CERN in the 1980s. The whole thing hangs together beautifully. But how did Glashow, Weinberg, and Salam know what to write down? How did they come to realize that "W and B mix up to produce the photon and the Z"? To answer that question is to catch a glimpse of the beautiful heart of modern particle physics. They did not simply guess, they had a big clue: Nature is symmetrical.

Symmetry is evident all around us. Catch a snowflake in your hand and look closely at this most beautiful of nature's sculptures. Its patterns repeat in a mathematically regular way, as if reflected in a mirror. More mundane, a ball looks unchanged as you turn it around, and a square can be flipped along its diagonal or along an axis that slices through its center without changing its appearance. In physics, symmetry manifests in much the same way. If we do something to an equation but the equation doesn't change, then the thing we did is said to be a symmetry of the equation. That's a little abstract, but remember that equations are the way physicists express how real things relate to one another. A simple but important symmetry possessed by all of the important equations in physics expresses the fact that if we pick up an experiment and put it on a moving train, then, provided the train isn't accelerating, the experiment will return the same results. This idea is familiar to us: It is Galileo's principle of relativity that lies at the heart of Einstein's theory. In the language of symmetry, the equations describing our experiment do not depend on whether the experiment is sitting on the station platform or onboard the train, so the act of moving the experiment is a symmetry of the equations. We have seen that this simple fact ultimately led Einstein to discover his theory of relativity. That is often the case: Simple symmetries can lead to profound consequences.

We're ready to talk about the symmetry that Glashow, Weinberg, and Salam exploited when they discovered the Standard Model of particle physics. The symmetry has a fancy name: gauge symmetry. So what is a gauge? Before we attempt to explain what it is, let's just say what it does for us. Let's imagine we are Glashow or Weinberg or Salam, scratching our heads as we look for a theory of how things interact with other things. We'll start by deciding we are going to build a theory of tiny, indivisible particles. Experiment has told us which particles exist, so we'd better have a theory that includes them all; otherwise, it will be only a half-baked theory. Of course, we could scratch our heads even more and try to figure out why those particular particles should be the ones that make up everything in the universe, or why they should be indivisible, but that would be a distraction. In fact, they are two very good questions to which we still do not have the answers. One of the qualities of a good scientist is to select which questions to ask in order to proceed, and which questions should be put aside for another day. So let's take the ingredients for granted and see if we can figure out how the particles interact with each other. If they did not interact with each other, then the world would be very boring—everything would pass through everything else, nothing would clump together, and we would never get nuclei, atoms, animals, or stars. But physics is so often about taking small steps, and it is not so hard to write down a theory of particles when they do not interact with each other—we just get the second line of the master equation with the *W*, *B*, and *G* bits scratched out. That's it—a quantum theory of everything but without any interactions. We have taken our first small step. Now here comes the magic. We shall demand that the world, and therefore our equation, have gauge symmetry. The consequence is astonishing: The remainder of the second line and the whole of the first line appear "for free." In other words, we are mandated to modify the "no interactions" version of the theory if we are to satisfy the demands of gauge symmetry. Suddenly we have gone from the most boring theory in the world to one in which the photon, *W*, *Z*, and gluon exist and, moreover, they are responsible for mediating all of the interactions between the particles. In other words, we have arrived at a theory that has the power to describe the structure of atoms, the shining of the stars, and ultimately the assembly of complex objects like human beings, all through the application of the concept of symmetry. We have arrived at the first two lines of our theory of nearly everything. All that remains is to explain what this miraculous symmetry actually is, and then those last two lines.

The symmetry of a snowflake is geometrical and you can see it with your eyes. The symmetry behind Galileo's principle of relativity isn't something you can see with your eyes, but it isn't too hard to comprehend even if it is abstract. Gauge symmetry is rather like Galileo's principle in that it is abstract, although with a little imagination it is not too hard to grasp. To help tie together the descriptions we offer and the mathematical underpinnings, we have been dipping into the master equation. Let's do it again. We said that the matter particles are represented by the Greek symbol ψ in the master equation. It's time now to delve just a little deeper. ψ is called a field. It could be the electron field, or an up-quark field, or indeed any of the matter particle fields in the Standard Model. Wherever it is biggest, that's where the particle is most likely to be. We'll focus on electrons for now, but the story runs just the same for all the other particles, from quarks to neutrinos. If the field is zero someplace, then the particle will not be found there. You might even want to imagine a real field, one with grass on it. Or perhaps a rolling landscape would be better, with hills and valleys. Where the hills are, the field is biggest, and in the valleys it is smallest. We are encouraging you to conjure up, in your mind's eye, an imaginary electron field. It might be surprising that our master equation is so noncommittal. It doesn't work with certainties and we cannot even track the electron around. All we can do is say that it is more likely to be found over here (where the mountain is) and less likely to be found over there (at base camp in the valley). We can put definite numbers on the chances of finding the electron to be here or there, but that is as good as it gets. This vagueness in our description of the world at the very smallest distance scales occurs because quantum theory reigns supreme there, and quantum theory deals only in the odds of things happening. There really does appear to be a fundamental uncertainty built into concepts such as position and momentum at tiny distances. Incidentally, Einstein really did not like the fact that the world should operate according to the laws of probability and it led him to utter his famous remark that "God does not play dice." Nevertheless, he had to accept that the quantum theory is extremely successful.

It explains all the experiments we have conducted in the sub-atomic world, and without it we would have no idea how the microchips inside a modern computer work. Maybe in the future someone will figure out an even better theory, but for now quantum theory constitutes our best effort. As we have been at pains to point out throughout this book, there is absolutely no reason why nature should work according to our common-sense rules when we venture to explain phenomena outside of our everyday experience. We evolved to be big-world mechanics, not quantum mechanics.

Returning to the task at hand, since quantum theory defines the rules of the game, we are obliged to talk of electron fields. But having specified our field and laid out the landscape, we are not quite done. The mathematics of quantum fields has a surprise lurking. There is some redundancy. For every point on the landscape, be it hill or valley, the mathematics says that we must specify not only the value of the field at a particular point (say, the height above sea level in our real-field analogy), corresponding to the probability that a particle will be found there, but we need also to specify something called the "phase" of the field. The simplest picture of a phase is to imagine a clock face or a dial (or a gauge) with only one clock hand. If the hand points to 12 o'clock, then that is one possible phase, or if it points to half-past, then that would be a different phase. We have to imagine placing a tiny clock face at each and every point on our landscape, with each one telling us the phase of the field at that point. Of course, these are not real clocks (and they certainly do not measure time). The existence of the phase is something that was familiar to quantum physicists well before Glashow, Weinberg, and Salam came along. More than that, everyone knew that although the relative phase between different points of the field matters, the actual value does not. For example, you could wind all of the tiny clocks forward by ten minutes and nothing would change. The key is that you must wind every clock by the same amount. If you forget to wind one of them, then you will be describing a different electron field. So there appears to be some redundancy in the mathematical description of the world.

Back in 1954, several years before Glashow, Weinberg, and Salam constructed the Standard Model, two physicists sharing an office at the Brookhaven Laboratory, Chen Ning Yang and Robert Mills, pondered the possible significance associated with the redundancy in setting the phase. Physics often proceeds when people play around with ideas without any good reason, and Yang and Mills did just that. They wondered what would happen if nature actually did not care about the phase at all. In other words, they played around with the mathematical equations while messing up all the phases, and tried to work out what the consequences might be. This might sound weird, but if you sit a couple of physicists in an office and allow them some freedom, this is the sort of thing they get up to. Returning to the landscape analogy, you might imagine walking over the field, haphazardly changing the little dials by different amounts. What happens is at first sight simple—you are not allowed to do it. It is not a symmetry of nature.

To be more specific, let's go back and look at only the second line of the master equation. Now strike out all of the W, B, and G bits. What we have is then the simplest possible theory of particles that we could imagine: The particles just sit around and never interact with each other. That little portion of the master equation very definitely does not stay the same if we suddenly go and redial all the little clocks (that isn't something that you are supposed to be able to see by just looking at the equation). Yang and Mills knew this, but they were more persistent. They asked a great question: How can we change the equation so that it *does* stay the same? The answer is fantastic. We need to add back precisely the missing bits of the master equation that we just struck out, and nothing else will do. In so doing we conjure into existence the force mediators and suddenly we go from a world without any interactions to a theory that has the potential to describe our real world. The fact that the master equation does not care about the values on the clock faces (or gauges) is what we mean by gauge symmetry. The remarkable thing is that demanding gauge symmetry leaves us no choice in what to write down: Gauge symmetry leads inexorably to the master equation. To put it another way, the forces that make our world interesting exist as a consequence of the fact that gauge symmetry is a symmetry of nature. As a postscript, we should add that Yang and Mills set the ball rolling, but their work was primarily of mathematical interest and it came well before particle physicists even knew which particles the fundamental theory ought to describe. It was Glashow, Weinberg, and Salam who had the wit to take their ideas and apply them to a description of the real world.

So we have seen how the first two lines of the master equation that underpins the Standard Model of particle physics can be written, and we hope to have given some flavor as to its scope and content. Moreover, we have seen that it is not ad hoc; instead

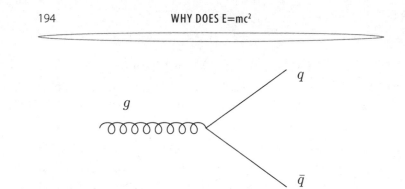

FIGURE 14

we are led inexorably to it by the draw of gauge symmetry. Now that we have a better feel for this most important of equations, we can get back to the task that originally motivated us. We were trying to understand to what extent nature's rules allow for the possibility that mass can actually be converted into energy, and vice versa. The answer lies, of course, within the master equation, for it spells out the rules of the game. But there is a much more appealing way to see what is going on and to understand how the particles interact with each other. This approach involves pictures, and it was introduced into physics by Richard Feynman.

What happens when two electrons come close to each other? Or two quarks? Or a neutrino gets close to an antimuon? And so on. What happens is that the particles interact with each other, in the precise way specified in the master equation. In the case of two electrons, they will push against each other because they have equal electric charge, whereas an electron and antielectron are attracted to each other because they have opposite electric charge. All of this physics resides in the first two lines of the master equation, and all of it can be summarized in just a

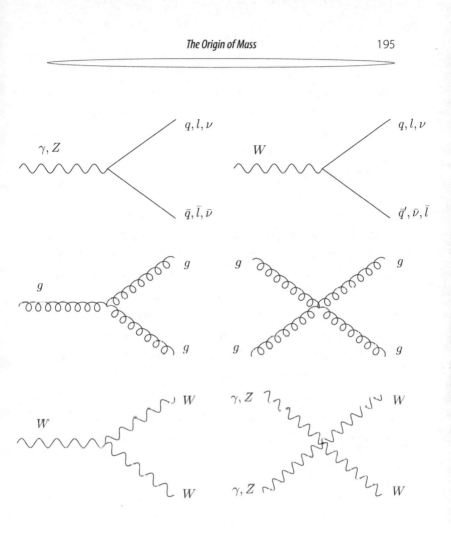

FIGURE 15

handful of rules that we can draw pictorially. It really is a very simple business to get a basic grasp of, although the details take a bit more effort to appreciate. We'll stick to the basics.

Looking again at the second line, the term that involves two ψ symbols and a G is the only portion of the equation that is relevant when quarks interact with each other via the strong force. Two quark fields and a gluon are interacting at the same point in spacetime—that is what the master equation is telling us. More than that, that is the *only* way they can interact with each other. That single portion of the master equation tells us how quarks and gluons interact, and it is prescribed precisely for us once we decide to make our theory gauge symmetric. We have absolutely no choice in the matter. Feynman appreciated that all of the basic interactions are this simple in essence, and he took to drawing pictures for each of the possible interactions that the theory allows. Figure 14 illustrates how particle physicists usually draw the quark-gluon interaction. The curly line represents a gluon and the straight line represents a quark or antiquark. Figure 15 illustrates the other allowed interactions in the Standard Model that come about from the first two lines of the master equation. Don't worry about the finer points of the pictures. The message is that we can write them down and that there aren't too many of them. Particles of light (photons) are represented by the symbol γ and the W and Z particles are labeled as such. The six quarks are labeled generically as q, the neutrinos appear as ν (pronounced "nu"), and the three electrically charged leptons (electron, muon, and tau) are labeled as l. Antiparticles are indicated by drawing a line over the corresponding symbol. Now here is the neat bit. These pictures rep-

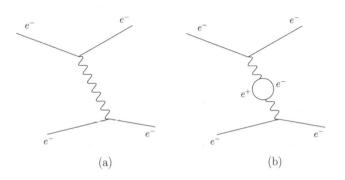

FIGURE 16

resent what particle physicists call interaction vertices. You are allowed to sew together these vertices into bigger diagrams, and any diagram you can draw represents a process that can happen in nature. Conversely, if you cannot draw a diagram, then the process cannot happen.

Feynman did a little more than just introduce the diagrams. He associated a mathematical rule with each vertex, and the rules are derived directly from the master equation. The rules multiply together in composite diagrams and allow physicists to calculate the likelihood that the process corresponding to a particular diagram will actually happen. For example, when two electrons encounter each other, the simplest diagram we can draw is as illustrated in Figure 16(a). We say the electrons scatter via the exchange of a photon. This diagram is built up by sewing together two electron-photon vertices. You should think of the two electrons heading in from the left, scattering off each other as a result of the photon exchange, and then heading out to the right. Actually, we have sneaked in another rule here. Namely, you are allowed to flip a particle to an antiparticle (and vice versa) provided you make it into an incoming particle. Figure 16(b) shows another possible way of sewing together the vertices. It is a little more fancy than the other figure, but again it corresponds to a possible way that the two electrons can interact. A moment's thought should convince you that there are an infinite number of possible diagrams. They all represent different ways that two electrons can scatter, but fortunately for those of us who have to calculate what is going on, some diagrams are more important than others. In fact, the rule is very easy to state: Generally speaking, the most important diagrams are the ones with the fewest vertices. So in the case of a pair of electrons, the diagram in Figure 16(a) is the most important one, because it has only two vertices. That means we can get a pretty good understanding of what happens by calculating only this diagram using Feynman's rules. It is delightful that what pops out of the math is the physics of how two electrically charged particles interact with each other, as discovered by Faraday and Maxwell. But now we can claim to have a much better understanding of the origin of this physics—we derived it starting from gauge symmetry. Calculations using Feynman's rules also give us much more than just another way to understand nineteenth-century physics. Even when two electrons interact, we can compute corrections to Maxwell's predictions—small corrections that improve upon his equations in that they agree better with the experimental data. So the master equation is breaking new ground. We really are just scratching the surface here. As we stressed, the Standard Model describes everything

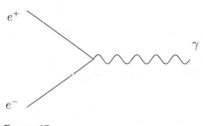

FIGURE 17

we know about the way particles interact with each other and it is a complete theory of the strong, weak, and electromagnetic forces, even succeeding in unifying two of them. Only gravity is excluded from this ambitious scheme to understand how everything in the universe interacts with everything else.

But we need to stay on message. How do Feynman's rules, which summarize the essential content of the Standard Model, dictate the ways in which we can destroy mass and convert it into energy? How can we use them to help us best exploit $E = mc^2$? First let us recall an important result from Chapter 5—light is made up of massless particles. In other words, photons do not have any mass. Now there is an interesting diagram we can draw—it is shown in Figure 17. An electron and an anti-electron (positron) bang into each other and annihilate to produce a single photon (for clarity we have labeled the electron e^- and the positron e^+). That is allowed by Feynman's rules. This diagram is noteworthy because it represents a case whereby we started with some mass (an electron and a positron have some mass) and we end up with no mass at all (a photon). It is the ultimate matter-destruction process, and all of the initial energy locked away inside the mass of the electron and antielectron is liberated as the energy of a photon. There is a hitch, though. The annihilation into a single photon is disallowed by the rule that everything that happens must simultaneously satisfy the laws of energy and momentum conservation, and this particular process cannot do that (it is not entirely obvious and we won't bother to prove it). It is a hitch that is easy to get around, though—make two photons. Figure 18 shows the relevant Feynman diagram—again, the initial mass is utterly destroyed and converted 100 percent into energy, in this case two photons. Processes like this played a very important role in the early history of the universe when matter and antimatter almost completely canceled themselves out by just such interactions. Today we see the remnant of that cancellation. Astronomers have observed that for every matter particle in the universe there are around 100 billion photons. In other words, for every 100 billion matter particles made just after the big bang, only one survived. The rest took the opportunity available to them, as pictured graphically in Feynman's diagrams, to divest themselves of their mass and become photons.

In a very real sense, the stuff of the universe that makes up stars, planets, and people is only a tiny residue, left over after the grand annihilation of mass that took place early on in the universe's history. It is very fortunate and almost miraculous that anything was left at all! To this day, we are not sure why that happened. The question "why is the universe not just filled with light and nothing else?" is still open-ended, and experiments around the world are geared up to help us figure out the answer. There is no shortage of clever ideas, but so far we have

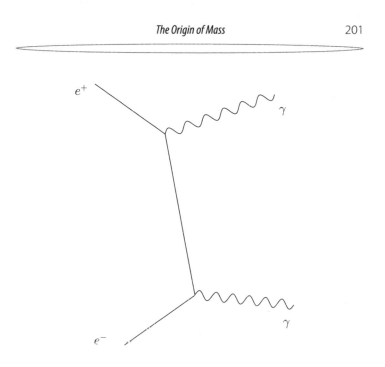

FIGURE 18

yet to find the decisive piece of experimental evidence, or proof that the theories are all wrong. The famous Russian dissident Andrei Sakharov carried out the pioneering work in this field. He was the first person to lay out the criteria that must be satisfied by any successful theory aiming to answer the question as to why there is any matter at all left over from the big bang.

We have learned that nature does have a mechanism for destroying mass, but unfortunately it is not very practical for use on Earth because we need a way of generating and storing antimatter—there is nowhere we can go to mine it and as far as we can tell, no lumps of it are lying around in outer space. As a fuel source it seems useless because there simply is no fuel. Antimatter can be

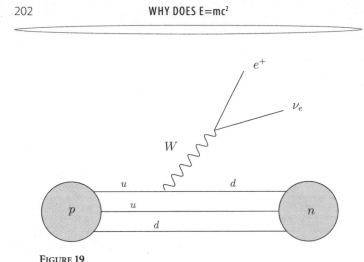

FIGURE 19

created in the laboratory, but only by feeding in lots of energy in the first place. So although the process of matter–antimatter annihilation represents the ultimate mechanism for converting mass to energy, it is not going to help us solve the world's energy crisis.

What about fusion, the process that powers the sun? How does that come about in the language of the Standard Model? The key is to focus our attention on the Feynman vertex involving a *W* particle. Figure 19 shows what is going on when a deuteron is manufactured from the fusion of two protons. Remember that protons are, to a good approximation, made up of three quarks: two up quarks and one down quark. The deuteron is made up of one proton and one neutron, and the neutron is again mainly made up of three quarks, but this time one up quark and two down quarks. The diagram shows how one of the protons can be converted into a neutron, and as you can see, the *W* particle is the key. One of the up quarks inside the proton has emitted a *W* particle and changed into a down quark as a result, thereby converting the proton into a neutron. According to the diagram, the *W* particle doesn't hang around. It dies and converts into an antielectron and a neutrino.* *W* particles emitted when a deuteron forms always die, and in fact nobody has ever seen *W* particles except via the stuff they turn into as they exit the world. As a rule of thumb almost all of the elementary particles die, because there is usually a Feynman vertex that allows it. The exception occurs whenever it is impossible to conserve energy or momentum, and that tends to mean that only the lightest particles stick around. That is the reason that protons, electrons, and photons dominate the stuff of the everyday world. They simply have nothing to decay into: The up and down quarks are the lightest quarks, the electron is the lightest charged lepton, and the photon has no mass. For example, the muon is pretty much identical to the electron except that it is heavier. Remember that we encountered it earlier when we were talking about the Brookhaven experiment. Since it starts out with more mass energy than an electron, its decay to an electron will not violate the conservation of energy. In addition, as illustrated in Figure 20, Feynman's rules allow it to happen and because a pair of neutrinos is also emitted there is no trouble conserving momentum. The upshot is that muons do decay and on average live for a fleeting 2.2 microseconds. Incidentally, 2.2 microseconds is a very long time on the

---

* Strictly speaking, it is an electron neutrino, because it is produced in conjunction with an antielectron.

timescale of most of the interesting particle physics processes. In contrast, the electron is the lightest Standard Model particle and it simply has nothing to decay into. As far as anyone can tell, an electron sitting on its own will never decay, and the only way to vanquish an electron is to make it annihilate with its antimatter partner.

Returning to the deuteron, Figure 19 explains how a deuteron can form from the collision of two protons, and it says we should expect to find one antielectron (positron) and one neutrino for every fusion event. As we have already mentioned, the neutrinos interact with the other particles in the universe only very weakly. The master equation tells us that is the case, for the neutrinos are the only particles that interact solely through the weak force. As a result, the neutrinos that are manufactured deep in the core of the sun can escape without too much trouble; they stream outward in all directions and some of them head out toward the earth. As with the sun, the earth is pretty much transparent to them and they pass through it without noticing it is even there. That said, each neutrino does have a very small chance of interacting with an atom in the earth, and experiments like Super-Kamiokande have detected them, as we discussed earlier.

How certain can we be that the Standard Model is correct, at least up to the accuracy of our current experimental capabilities? Over many years now the Standard Model has been put through the most rigorous tests at various laboratories around the world. We don't need to worry that the scientists are biased in favor of the theory; those conducting the tests would dearly love to find that the Standard Model is broken or deficient in

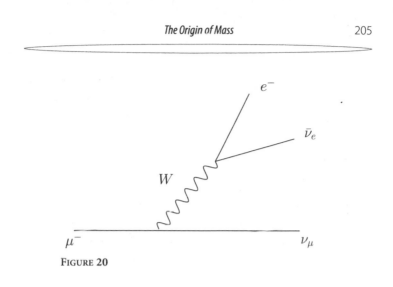

FIGURE 20

some way, and they are trying hard to test it to destruction. Catching a glimpse of new physical processes, which may open up dazzling new vistas with magnificent views of the inner workings of the universe, is their dream. So far the Standard Model has withstood every test.

The most recent of the big machines used to test it is the Large Hadron Collider (LHC) at CERN. This worldwide collaboration of scientists aims to either confirm or break the Standard Model; we shall return to the LHC shortly. The predecessor to the LHC was the Large Electron Positron Collider (LEP), and it succeeded in delivering some of the most exquisite tests to date. LEP was housed inside a 27-kilometer circular tunnel running underneath Geneva and some picturesque French villages, and it explored the world of the Standard Model for eleven years, from 1989 until 2000. Large electric fields were used to accelerate beams of electrons in one direction and of positrons in the other. Crudely speaking, the acceleration of charged particles by electric fields is similar to the mechanism used to shoot electrons at old-fashioned CRT (cathode ray tube) television screens to produce the picture. The electrons are emitted at the back of the set, and that is why older TVs tend to be quite bulky. Then the electrons are accelerated by an electric field to the screen at the front of the TV. A magnet makes the beam bend and scan across the screen to make the picture.

At LEP, magnetic fields were also exploited, this time to bend the particles in a circle so they followed the arc of the tunnel. The whole point of the venture was to bring the two beams of particles together so they would collide head-on. As we have already learned, the collision of an electron and a positron can lead to the annihilation of both, with their mass converting into energy. This energy is what physicists at LEP were most interested in, because it could be converted into heavier particles in accord with Feynman's rules. During the first phase of the machine's operation, the electron and positron had energies that were very precisely tuned to the value that greatly enhanced the chances of making a Z particle (you might want to check back to the list of Feynman's rules in the Standard Model to check that electron-positron annihilation into a Z particle is allowed). The Z particle is actually pretty heavy by the standards of the other particles—it is nearly 100 times more massive than a proton and nearly 200,000 times more massive than the electron and positron. As a result, the electron and positron had to be pushed to within a whisker of the speed of light to have energy sufficient to bring the Z into being. Certainly the energy locked in their mass and liberated upon annihilation is nowhere near sufficient to make the Z.

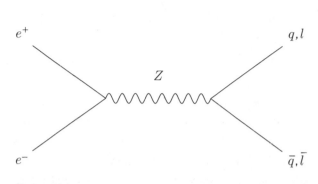

e^+ q, l

Z

e^- $\bar{q}, \bar{l}$

FIGURE 21

The initial goal of LEP was simple: keep on producing Z particles by repeatedly colliding electrons and positrons. Every time the particle beams collide, there would be a reasonable chance of an electron in one beam annihilating against a single positron in the other beam, resulting in the production of a single Z particle. By quick-firing beams into each other, LEP managed to make over 20 million Z particles through electron-positron annihilation during its lifetime.

Just like the other heavy Standard Model particles, the Z is not stable and it lasts for a fleeting 10^{-25} seconds before it dies. Figure 21 illustrates the various possible Z particle processes that the 1,500 or so LEP physicists were so interested in, not to mention the many thousands more around the world who were eagerly awaiting their results. Using giant particle detectors that surround the point where the electron and positron annihilate each other, particle physicists could capture the stuff produced by the decay of the Z and identify it. Modern particle physics detectors, like those used at LEP, are a little like huge digital cameras, many meters across and many meters tall, that can track particles as they pass through them. They, like the accelerators themselves, are glorious feats of modern engineering. In caverns as big as cathedrals, they can measure a single subatomic particle's energy and momentum with exquisite accuracy. They are truly at the edge of our engineering capabilities, which makes them wonderful monuments to our collective desire to explore the workings of the universe.

Armed with these detectors and vast banks of high-performance computers, one of the primary goals for the scientists involved a pretty simple strategy. They needed to sift through their data to identify those collisions in which a Z particle was produced and then for each collision, figure out how the Z particle decayed. Sometimes it would decay to produce an electron-positron pair; other times a quark and antiquark would be produced or maybe a muon and an antimuon (see Figure 21 again). Their job was to keep a tally of how many times the Z decayed through each of the possible mechanisms predicted by the Standard Model and compare the results with the expected numbers as predicted by the theory. With over 20 million Z particles on hand, they could make a pretty stringent test of the correctness of the Standard Model and, of course, the evidence showed that the theory works beautifully. This exercise is called measuring the partial widths, and it was one of the most important tests of the Standard Model that LEP provided. Over time, many other tests were performed and in all cases the Standard Model theory was seen to work. When LEP was finally shut down in 2000, its ultraprecise data had been able to test the Standard Model to a precision of 0.1 percent.

Before we leave the subject of testing the Standard Model, we cannot resist one other example from a quite different type of experiment. Electrons (and many other elementary particles) behave like tiny magnets, and some very beautiful experiments have been designed to measure these magnetic effects. These aren't collider experiments. There is no brutal smashing together of matter and antimatter here. Instead, very clever experiments allow the scientists to measure the magnetism to an astonishing one part per trillion. It is a staggering precision, akin to measuring the distance from London to New York to an accuracy much less than the thickness of a human hair. As if that weren't amazing enough, the theoretical physicists have been hard at work too. They have calculated the same thing. Calculations like this used to be done using nothing more than a pen and some paper, but these days even the theorists need good computers.

Nevertheless, starting with the Standard Model and a cool head, theorists have calculated the predictions of the Standard Model, and their result agrees exactly with the experimental number. To this day the theory and experiment are in agreement to ten parts per billion. It is one of the most precise tests of any theory that has ever been made in all of science. By now, and thanks in no small part to LEP and the electron magnetism experiments, we have a great deal of confidence that the Standard Model of particle physics is on the right lines. Our theory of nearly everything is in fine shape—except for one last detail, which is actually a fairly big detail. What are those last two lines of the master equation?

We are guilty in fact of hiding one crucial piece of information that is absolutely central to our quest in this book. Now is the time to let the cat out of the bag. The requirement of gauge symmetry seems to demand that all of the particles in the Standard Model have no mass. That is plain wrong. Things do have mass and you do not need a complicated scientific experiment to prove it. We've spent the entire book so far thinking about it, and we derived the most famous equation in physics, $E = mc^2$, and that very definitely has an "m" in it. The final two lines of the master equation are there to fix this problem. In understanding those final two lines we will complete our journey, for we will have an explanation for the very origin of mass.

The problem of mass is very easy to state. If we try to add mass directly into the master equation, then we are doomed to spoil gauge symmetry. But as we have seen, gauge symmetry lies at the very heart of the theory. Using it, we were able to conjure into being all of the forces in nature. Worse still, theorists proved in the 1970s that abandoning gauge symmetry is not an option, because then the theory falls apart and stops making sense. This apparent impasse was solved by three groups of people working independently of each other in 1964. François Englert and Robert Brout working in Belgium, Gerald Guralnik, Carl Hagen, and Tom Kibble in London, and Peter Higgs in Edinburgh all wrote landmark papers that led to what later became known as the Higgs mechanism.

What would constitute an explanation of mass? Well, suppose you started out with a theory of nature in which mass never reared its head. In such a theory, mass simply does not exist and you would never invent a word for it. As we have learned, everything would whiz around at the speed of light. Now, suppose that within that theory something happens such that after the event the various particles start to move around with different, slower speeds and certainly no longer move at light speed. Well, you would be quite entitled to say that the thing that happened is responsible for the origin of mass. That "thing" is the Higgs mechanism, and now is the time to explain what it is.

Imagine you are blindfolded, holding a ping-pong ball by a thread. Jerk the string and you will conclude that something with not much mass is on the end of it. Now suppose that instead of bobbing freely, the ping-pong ball is immersed in thick maple syrup. This time if you jerk the thread you will encounter more resistance, and you might reasonably presume that the thing on the end of the thread is much heavier than a ping-pong ball. It is as if the ball is heavier because it gets dragged back by the syrup. Now imagine a sort of cosmic maple syrup that pervades the whole of space. Every nook and cranny is filled with it, and it is so pervasive that we do not even know it is there. In a sense, it provides the backdrop to everything that happens.

The syrup analogy only goes so far, of course. For one thing, it has to be selective syrup, holding back quarks and leptons but allowing photons to pass through it unimpeded. You might imagine pushing the analogy even further to accommodate that, but we think the point has been made and we ought not forget that it is an analogy, after all. The papers of Higgs et al. certainly never mention syrup.

What they do mention is what we now call the Higgs field. Just like the electron field, it has associated with it a particle: the Higgs particle. Just like the electron field, the Higgs field fluctuates, and where it is biggest the Higgs particle is more likely to

be found. There is a big difference, though: The Higgs field is
not zero even when no Higgs particles are around, and that is
the sense in which it is like all-pervasive syrup. All of the particles in the Standard Model are moving around in the background of the Higgs field, and some of the particles are affected
by it more than others. The last two lines of the master equation
capture just this physics. The Higgs field is represented by the
symbol ϕ and the portions of the third line that involve two in-
stances of ϕ along with a B or a W (which in our compressed
notation are tucked away inside the D symbol in the third line
of the master equation) are the terms that generate masses for
the W and Z particles. The theory is cleverly arranged so the
photon remains massless (the piece of the photon that sits in B
and the piece in W cancel out in the third line; again, that's all
hidden in the D symbol) and since the gluon field (G) never appeared, it too has no mass. That is all in accord with experiment.
Adding the Higgs field has generated masses for the particles
and it has done so without spoiling the gauge symmetry. The
masses are instead generated as a result of an interaction with
the background Higgs field. That is the magic of the whole
idea—we can get masses for the particles without paying the
price of losing gauge symmetry. The fourth line of the master
equation is the place where the Higgs field generates the masses
for the remaining matter particles of the Standard Model.

There is a snag to this fantastic picture. No experiment has
ever seen a Higgs particle. Every other particle in the Standard
Model has been produced in experiments, so the Higgs really
is the missing piece in the entire jigsaw. If it does exist as predicted, then the Standard Model will have triumphed again,

and it can add an explanation for the origin of mass to its impressive list of successes. Just like all the other particle interactions, the Standard Model specifies exactly how the Higgs particle should manifest itself in experiments. The only thing it doesn't tell us is how heavy it is, although it does predict that the Higgs mass should lie within a particular range now that we know the masses of the W particle and the top quark. LEP could have seen the Higgs if it had been at the lighter end of the predicted range, but since none were seen, we might presume it is too heavy to have been produced at LEP (remember that heavier particles need more energy to produce them, by virtue of $E = mc^2$). At the time of writing, the Tevatron collider at the Fermi National Accelerator Laboratory (Fermilab) near Chicago is hunting for the Higgs, but again it has not to date seen a hint. It is again very possible that the Tevatron has insufficient energy to deliver a clear Higgs signal, although it is very much in the race. The LHC is the highest-energy machine ever built, and it really should settle the question of the Higgs's existence once and for all because it has enough energy to reach well beyond the upper limits set by the Standard Model. In other words, the LHC will either confirm or break the Standard Model. We'll return shortly to explain why we are so sure that the LHC will do the job the earlier machines have failed to do, but first we would like to explain just how the LHC expects to make Higgs particles.

The LHC was built within the same 27-kilometer-circumference tunnel that LEP used but, apart from the tunnel, everything else has changed. An entirely new accelerator now occupies the space LEP once occupied. It is capable of accelerating

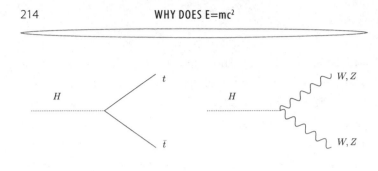

FIGURE 22

protons in opposite directions around the tunnel to an energy equal to more than 7,000 times their mass energy. Smashing the protons into each other at these energies advances particle physics into a new era, and if the Standard Model is right, it will produce Higgs particles in large numbers. Protons are made up of quarks, so if we want to figure out what should happen at the LHC, then all we need to do is identify the relevant Feynman diagrams.

The most important vertices corresponding to interactions between the regular Standard Model particles and the Higgs particle are illustrated in Figure 22, which shows the Higgs as a dotted line interacting with the heaviest quark, the top quark (labeled t), and with the also pretty heavy W or Z particles. Perhaps it will come as no surprise that the particle responsible for the origin of mass prefers to interact with the most massive particles around. Knowing that the protons furnish us with a source of quarks, our task is to figure out how to embed the Higgs vertex into a bigger Feynman diagram. Then we'll have figured out how Higgs particles can be manufactured at the LHC. Since quarks interact with W (or Z) bosons, it is easy to work out how the Higgs could be produced via W (or Z) particles. The result

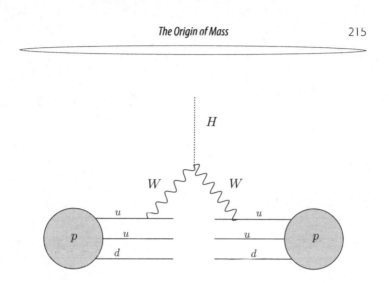

FIGURE 23

is shown in Figure 23: A quark from each of the colliding protons (labeled "*p*") emits a *W* (or *Z*) particle, and these fuse together to make the Higgs. The process is called weak boson fusion, and it is expected to be a key process at the LHC.

The case of the top quark production mechanism is a little trickier. Top quarks do not exist inside protons, so we need a way to go from the light (up or down) quarks to top quarks. Well, top quarks interact with the lighter quarks through the strong force—i.e., mediated by emitting and absorbing a gluon. The result is shown in Figure 24. It is rather similar to the weak boson fusion process except that the gluons replace the *W* or *Z*. In fact, because this process proceeds through the strong force, it is the most likely way to produce Higgs particles at the LHC. It goes by the name of gluon fusion.

This then is the Higgs mechanism, the currently most widely accepted theory for the origin of mass in the universe. If all goes

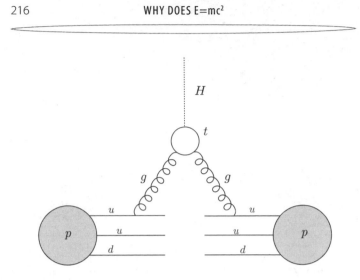

FIGURE 24

according to plan, the LHC will either confirm the Standard
Model description of the origin of mass or show that it is wrong.
This is what makes the next few years such an exciting time for
physics. We are in the classic scientific position of having a
theory that predicts precisely what should happen in an experiment, and will therefore stand or fall depending on the results
of that experiment. But what if the Standard Model is wrong?
Couldn't something totally different and unexpected happen?
Maybe the Standard Model is not quite right and there is no
Higgs particle. There is no arguing that these are genuine possibilities. Particle physicists are particularly excited because they
know that the LHC *must* reveal something new. The possibility
that the LHC will see nothing new is not an option at all because the Standard Model, stripped of the Higgs, just does not
make sense at the energies that the LHC is capable of generat-

ing, and the predictions of the Standard Model simply fall apart. The LHC is the first machine to enter this uncharted area. More specifically, when two *W* particles collide at energies in excess of 1,000 times the proton's mass energy, as they certainly will at the LHC, we lose the ability to calculate what is happening if we simply throw the Higgs parts of the master equation away. Adding the Higgs back in makes the calculations work out, but there are other ways to make the *W* scattering process work—and the Higgs is not the only option. Whichever way nature chooses, it is absolutely unavoidable that the LHC will measure something that necessarily contains physics we have never encountered before. It is not common for scientists to perform an experiment with such a guarantee that interesting things are going to reveal themselves, and this is what makes the LHC the most eagerly anticipated experiment in many years.

8

Warping Spacetime

Thus far we have thought of spacetime as fixed and unchanging—something akin to a four-dimensional stage or the arena within which "things happen." We have also come to appreciate that spacetime has a geometry and that the geometry is most certainly not that of Euclid. We have seen how the idea of spacetime leads naturally to $E = mc^2$ and how this simple equation and the physics it represents has become a foundation stone of both our modern theories of nature and the industrial world. Let us move toward the final twist in our story by asking one last curiosity-driven question: Is it possible that spacetime could be warped and curved differently from place to place in the universe?

The idea of curved space should not be new to us, of course. Euclidean space is flat and Minkowski space is curved. By which we mean that Pythagoras' theorem doesn't apply in Minkowski spacetime. Instead, the minus-sign version of the distance equation applies. We also know that the distance between two points in spacetime is analogous to the distance between different places on a map of the earth, in that the shortest distance between two points is not a straight line in the usual sense of the word. So Minkowski spacetime and the surface of the earth are examples of curved spaces. Having said that, the distance between two points in Minkowski spacetime does always satisfy $s^2 = (ct)^2 - x^2$, and this means that it curves in the same way everywhere. The same can be said for the surface of the earth. Might it, however, make sense to speak of a surface that curves differently from place to place? What would spacetime look like if this were allowed, and what would the implications be for clocks, rulers, and the laws of physics? To explore this admittedly rather arcane-sounding possibility, we shall once again take a step down from the mind-bending four dimensions to the commonplace two dimensions and focus our attention on the surface of a sphere.

A smooth ball is curved the same way everywhere—that much is obvious. But a golf ball, with dimples in it, is not. Likewise, the earth's surface is not a perfect sphere. As we zoom in, we see valleys and hills, mountains and oceans. The law for the distance between two points on the earth's surface is only approximately the same everywhere. For a more precise answer we need to know how the earth's undulating surface changes as we journey over the mountains and through the valleys between the start and finish of any journey. Could spacetime have dimples in it like a golf ball or mountains and valleys like the earth? Might it "warp" from place to place?

When we first derived the distance equation in spacetime, it seemed that we had no flexibility to change it from place to place. Indeed we argued that the precise form of the distance equation was forced upon us by the constraints of causality. But we did make a very big assumption. We *assumed* that spacetime is the same everywhere. It is true enough to say that this turns out to be an assumption that works remarkably well and the experimental evidence is largely in its favor, for this assumption was a crucial one on the road to $E = mc^2$. But maybe we have not looked carefully enough. Might spacetime not be the same everywhere, and might this lead to consequences that we can observe? The answer is emphatically yes. To arrive at this conclusion, let us follow Einstein on one last journey. It was a journey that caused him ten years of hard struggle before he finally arrived at yet another majestic destination: the theory of general relativity.

Einstein's journey to special relativity was triggered by a simple question—what would it mean if the speed of light were the same for all observers? His rather more tortuous journey to general relativity began with an equally simple observation that impressed him so much that he could not rest until he had recognized its true significance. The fact is this: All things fall to the ground with the same acceleration. That's it . . . that is what excited Einstein so much! It takes a mind like Einstein's to recognize that such an apparently benign fact could be of very deep significance.

Actually, this is a famous result in physics, known long before Einstein came along. Galileo is credited with being the first to recognize it. Legend has it that he climbed up the Leaning Tower of Pisa, dropped two balls of different masses off the top, and observed that they hit the ground at the same time. Whether he actually carried out the experiment does not really matter; what is important is that he correctly recognized what the outcome would be. We do know for sure that the experiment was eventually performed, not in Pisa but on the moon in 1971 by Apollo 15 commander David Scott. He dropped a feather and a hammer and both hit the ground at the same time. We can't do that experiment on earth because a feather gets caught by the wind and slows down, but it is quite spectacular when performed in the high vacuum of the lunar surface. There isn't much need to go all the way to the moon to check that Galileo was right, of course, but that doesn't detract from the drama of the Apollo 15 demonstration, and the video is well worth watching. The important fact is that everything falls at the same rate, if complicating factors such as air resistance can be removed. The obvious question is why? Why do they fall at the same rate, and why are we making it out to be such a big deal?

Imagine you are standing in a stationary elevator. Your feet press firmly on the ground and your head pushes down on your shoulders. Your stomach rests in place inside your body. Now imagine you have the misfortune to be inside an elevator that is plummeting toward the ground because the cables have been cut. Since everything falls at the same rate, your feet no longer push onto the floor of the lift, your head no longer pushes onto your shoulders, and your stomach floats freely inside your body. In short, you are weightless. This is a big deal because it is *exactly* as if someone had turned off gravity. An astronaut floating freely in outer space would feel just the same. To be a little more precise, as the lift falls there are no experiments that you can do inside the lift that are able to distinguish between the possibilities that you are plummeting toward earth or floating in outer space. Of course you know the answer because you walked into the elevator, and perhaps the floor counter is whizzing toward "ground" at an alarming rate, but that is not the point. The point is that the laws of physics are identical in the two cases. That is what affected Einstein so deeply. The universality of free fall has a name. It is called the principle of equivalence.

Generally speaking, gravity changes from place to place. Its pull is stronger the closer to the center of the earth you are, although there isn't that much difference between sea level and the top of Mount Everest. It is much weaker on the moon, because the moon is less massive than the earth. Likewise, the gravitational pull of the sun is much stronger than that of the earth. But wherever you happen to be in our solar system, the force of gravity will not vary too much within your immediate locality. Imagine standing on the ground. The gravity at your feet will be slightly stronger than the gravity at your head but it will be a very small difference. It will be smaller for a short person and bigger for a tall person. You might imagine a tiny ant. The difference in the gravitational pull on its feet compared to its head will be smaller still. Let's travel the well-worn pathways of the thought experiment one more time and imagine smaller and smaller things, all the way down to a tiny "elevator." So small is our elevator that the gravity can be assumed to be the same everywhere inside it. The tiny elevator is populated by even tinier physicists whose job it is to carry out scientific experiments within their elevator. Now we can imagine that the little elevator is in free fall. In this case, none of the tiny physicists would ever utter the word "gravity." A description of the world in terms of observations made by this group of tiny falling physicists has the astonishing virtue that gravity simply does not exist. Nobody would utter the word "gravity" in their tiny squeaky voices because there is no observation that could be made within the elevators that would indicate that there was such a thing. But hang on a second! Clearly something makes the earth orbit the sun. Is this just some clever sleight of hand or are we onto something important?

Let's leave gravity and spacetime for a moment and return to the analogy of the curved surface of the earth. A pilot planning a trip from Manchester to New York clearly needs to recognize that the earth's surface is curved. In contrast, when moving between your dining room and your kitchen you can safely ignore the curvature of the earth and assume that the surface is flat. In other words, the geometry is (very nearly) Euclidean. This is ultimately why it took awhile for humans to discover that the earth is not flat but spherical; the radius of curvature is very much bigger than the day-to-day distances that we are used to dealing with. Let's imagine chopping up the earth's surface into lots of little square patches, as illustrated in Figure 25. Each patch is pretty near flat, and the smaller we make the patches, the nearer to flat each one is. On each patch, Euclid's geometry holds sway: Parallel lines don't cross and Pythagoras' theorem works. The curvature of the surface becomes evident only when we try to cover large areas of the earth's surface with our Euclidean patches. We need *lots* of little patches sewn together to faithfully construct the curved surface of the sphere.

Now let's return to our little elevator in free fall and imagine it is accompanied by many other little elevators, one at each point in spacetime, in fact. The spacetime inside each is approximately the same everywhere, and the approximation gets

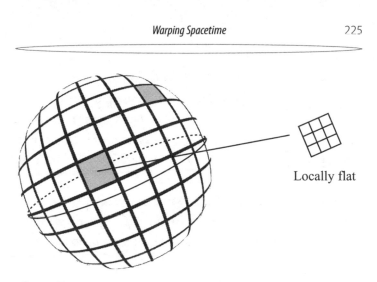

Locally flat

FIGURE 25

better as the elevators get smaller. Now, recall that in Chapter 4 we were very careful to point out our assumption that spacetime should be "unchanging and the same everywhere," and this was critical in allowing us to construct Minkowski's spacetime distance formula. Since the spacetime within each tiny elevator is also "unchanging and the same everywhere," it therefore follows that we can use Minkowski's distance formula inside each individual little elevator.

Hopefully, the analogy with the sphere is beginning to emerge. For "flat patch on the earth's surface," read "falling elevator in spacetime," and for "curved surface of the earth," read "curved spacetime." In fact, physicists often refer to Minkowski spacetime as "flat spacetime" for this very reason. Minkowski spacetime plays the role of flat Euclidean space in the analogy. In this book, we've reserved the use of the word "flat" for Euclidean geometry, and the minus sign in the Minkowskian version of

Pythagoras' theorem motivated us to use the term "curved." Sometimes the use of language is not as straightforward as we might like it to be! So the assembly of little elevators is to space-time as the assembly of little patches is to the sphere. In each little elevator, gravity has been banished, but we could imagine sewing all the little Minkowski patches together to form a curved spacetime in exactly the same way that we constructed the curved surface of the earth from flat Euclidean patches. If there were no gravity, then we could get by with one big eleva-tor within which the geometry is that of Minkowski. So what we have just learned is that if there is gravity around, we can make it go away but only at the expense of making spacetime curved. What a remarkable conclusion.

Turn this around, and it looks like we have discovered that the force of gravity is actually nothing more than a signal to us that spacetime itself is curved. Is this really true, and what causes the curving? Since gravity is found in the vicinity of matter, we might conclude that spacetime is warped in the vicinity of matter and, since $E = mc^2$, energy. The amount of warping is something we have so far said nothing at all about. And we don't intend to say very much because it is, to use a well-worn physics phrase, nontrivial. In 1915, Einstein wrote down an equation that was able to quantify exactly how much warping there should be in the presence of matter and energy. His equation improves upon Newton's age-old law of gravity in that it is automatically in accord with the special theory of relativity (Newton's law is not). Of course, it gives very similar results to Newton's theory for most cases we encounter in everyday life, but it does expose Newton's theory as an approximation. To illustrate the different ways of thinking about gravity, let's see how Newton and Einstein would describe the way in which the earth orbits the sun. Newton would say something like this: "The earth is pulled toward the sun by the force of gravity, and that pull prevents it from flying off into space, constraining it instead to move in a big circle."* It is similar to whirling a ball on a string around your head. The ball will follow a circular path because the tension in the string prevents it from doing otherwise. If you cut the string, the ball would head off in a straight line. Likewise, if you suddenly turned off the sun's gravity, Newton would say that the earth would then head off into outer space in a straight line. Einstein's description is quite different and goes like this: "The sun is a massive object and as such it distorts spacetime in its vicinity. The earth is moving freely through spacetime but the warping of spacetime makes the earth go in circles."

To see how an apparent force might be nothing more than a consequence of geometry, we can consider two friends walking on the earth's surface. They are told to begin at the equator and to walk due north parallel to each other in perfect straight lines, which they dutifully do. After a while, they will notice that they are coming closer together and, if they carry on walking for long enough, they will bump into each other at the North Pole. Having established that neither of them cheated and wandered off course, they may well conclude that a force acted between them that pulled them together as they walked northward. This is one

---

* Actually, it moves in an ellipse, a slightly squashed circle, but it is pretty close to a circle.

way to think about things, but there is of course another explanation: The surface of the earth is curved. The earth is doing much the same thing as it moves around the sun.

To get a better feel for what we are talking about, let's return to one of our intrepid walkers on the surface of the earth. As before, he is told always to walk in a straight line. Locally, that is an instruction he can follow without any confusion because at any point on the earth he can assume Euclidean geometry works just fine and, as a result, the idea of a straight line is clear to him. Even so, he ends up walking in a circular path, although we can think of the circle as being build up of lots of little straight lines. Now let's return to the case of gravity and spacetime. The notion of straight lines through curved spacetime is entirely analogous to the notion of straight lines on the earth's surface. The complication arises because spacetime is a four-dimensional "surface," while the earth's surface is only two-dimensional. But once again the complication is more to do with our limited imagination rather than any increase in mathematical complexity. In fact, the mathematics of geometry on the surface of a sphere is no harder than the mathematics of geometry in spacetime. Armed with the idea of straight lines (they are also known as geodesics) in spacetime we might be so bold as to suggest how gravity works. We have seen that gravity can be banished in exchange for curved spacetime and that locally the spacetime is the "flat" spacetime of Minkowski. We know very well by this point in the book how things move in such an environment. For example, if a particle is at rest it will remain so (unless something comes along and gives it a push or pull). That means it follows a spacetime trajectory that moves only along the time axis. Likewise, objects that are moving with a constant speed will carry on moving in the same direction and at the same speed (again, unless something comes and knocks them off course). In this case they will follow straight lines on the spacetime diagram that are tilted away from the time axis So, on each tiny patch of spacetime everything should follow a straight line unless acted upon by some external influence. The whole appearance of gravity emerges when we sew all of the little patches together; for only then do the individual straight lines join together into something more interesting, like the orbit of a planet around the sun. We have not said how to join up the patches in order to build the warping of spacetime, and it is Einstein's equation of 1915 that determines exactly how we are to do that. But the bottom line could not be much simpler—gravity has been banished in exchange for pure geometry.

So gravity is geometry and all things move along straight lines in spacetime unless they are knocked off course. But at any given point in spacetime there is an infinite number of geodesics, just as there is an infinite number of straight lines passing through any point on the earth's surface (or any other surface, for that matter). So how are we to figure out which spacetime trajectory an object will move along? The answer is simple enough: Circumstances dictate it. For example, the person on the trek around the earth could start out in any number of directions. He chooses which route to take. Likewise, an object dropped from rest near to the earth will start out on one spacetime geodesic while one that is thrown will start out on a different geodesic. By specifying the direction an object moves through spacetime at any particular point, we therefore know its complete trajectory.

Moreover, all objects heading off in that particular direction necessarily follow the same trajectory, irrespective of their internal properties (like mass or electric charge). They just follow a straight line, and that's all there is to it. In this way the curved spacetime view of gravity beautifully expresses the principle of equivalence that so captivated Einstein.

Our musings on the nature of space and time have led us to understand that the earth is doing nothing more than falling in a straight line around the sun. It is just that the straight line is in a curved spacetime, which manifests itself as a (nearly) circular orbit in space. We have not gone ahead and proved that the sun warps spacetime such that the earth falls along a geodesic whose shadow in three-dimensional space is (nearly) a circle. We haven't done it simply because it involves too much mathematics. It also involves us making some statement as to how objects actually warp spacetime, and we have been ducking that issue. The mathematical complexity is the main reason why it took Einstein ten years to develop the theory. General relativity is conceptually rather simple but mathematically difficult, although the difficulty most definitely does not obscure its beauty. Indeed many physicists consider Einstein's theory of general relativity to be the most beautiful of all our theories of nature.

You may well have noticed that nothing we have said has singled out one type of object over another. In particular, light itself should also move through spacetime along a geodesic. In each spacetime patch that it passes over, the light travels along one of the 45-degree lines we introduced in Chapter 4 but, upon sewing all the patches together, we will find a trajectory that bends through space. The bending simply reflects the way in which the spacetime is warped by the presence of mass and energy. Just as for the case of the earth in orbit around the sun, its path through space is a shadow of its four-dimensional geodesic. The power of the equivalence principle and the implied bending of light can be illustrated nicely by another thought experiment.

Imagine that you are standing on the earth and you fire a laser beam horizontally. What happens to it? The principle of equivalence tells us what happens. The light falls toward the ground at exactly the same rate as would an object that is released from rest at the precise moment that the laser is fired. If Galileo had access to a laser and he fired it horizontally off the Leaning Tower of Pisa at the same time as dropping a cannonball, then Einstein predicts that the laser beam would hit the ground at the same time as the cannonball. The problem with this experiment in reality is that the earth's surface curves away very quickly and the laser would never actually hit the ground because it would run out of earth. If we imagine instead that we are standing on a flat earth, then that problem goes away and we would expect the laser beam to hit the ground at exactly the same time as the cannonball, only a very great deal farther away. In fact, if the cannonball took a second to hit the ground, then the laser would hit the ground one light-second from the tower, which is just over 186,000 miles away.

The description of gravity as geometry is certainly immensely satisfying and it leads to quite startling conclusions but, as we have emphasized throughout this book, it is ultimately useless unless it leads to predictions that can be tested against experiment. Fortunately for Einstein, he had to wait only four years for his exotic predictions to be confirmed.

The first great test of Einstein's new theory came in 1919 when Arthur Eddington, Frank Dyson, and Charles Davidson wrote a paper titled "A Determination of the Deflection of Light by the Sun's Gravitational Field, from Observations Made at the Total Eclipse of May 29, 1919." The paper was published in the *Philosophical Transactions of the Royal Society of London* and contains the immortal words "both of these point to the full deflection of 1."75 of Einstein's generalized relativity theory." Overnight, Einstein became a global superstar. His esoteric theory of curved spacetime had been vindicated following the not inconsiderable efforts of Eddington, Dyson, and Davidson: To see the eclipse, they had to make expeditions to Sobral in Brazil and Principe, off the western coast of Africa. The eclipse allowed them to look at stars lurking very close to the sun that would otherwise be obscured by its light. This is the starlight best suited to testing Einstein's theory, because it should be deflected the most since the spacetime curvature is greater the closer you get to the sun. In essence, Eddington, Dyson, and Davidson were looking to see whether the stars shifted their position in the sky as the sun passed by. Quite literally, the sun bends spacetime and acts like a lens, distorting the pattern of stars on the sky.

Today Einstein's theory has been tested to a high accuracy using some of the most remarkable objects in the universe: spinning neutron stars called pulsars. We met neutron stars and pulsars at the end of Chapter 6, and they are abundant in the universe. Of all the objects we can study accurately from the earth using telescopes, spinning neutron stars are special in that they provide us with large distortions of spacetime and a precise time stamp that rivals the stability of the world's best atomic clocks. If you wanted to dream up an object that would provide the perfect environment in which to test general relativity, you might well come up with a pulsar. Pulsars deliver their time stamp by beaming out radio waves as they spin. You might like to imagine a lighthouse, shining out a narrow beam that scans around once every second or so. These wonderfully useful objects were discovered quite by accident in 1967 by Jocelyn Bell Burnell and Tony Hewish. If you're wondering how it is possible to stumble across a spinning neutron star by accident, Bell Burnell was looking for fluctuations in the intensity of radio waves emitted by distant objects known as quasars. The fluctuations were known to be caused by the solar winds in interstellar space. Being a good scientist, however, she was always on the lookout for interesting things in her data and, one November night, she detected a regular signal that she and her supervisor, Hewish, naturally thought was of man-made origin. Subsequent observations convinced them that this could not be the case and that the signal must come from a source beyond our planet. "I went home that evening very cross," Bell Burnell later said of her observations. "Here was I trying to get a PhD out of a new technique, and some silly lot of little green men had to choose my aerial and my frequency to communicate with us."

Although pulsars are fairly commonplace in the universe, there is only one known instance where two pulsars are circling each other. The existence of this double pulsar was established by radio astronomers in 2004, and subsequent observations have led to the most precise test to date of Einstein's general theory.

The double pulsar is a remarkable thing. We now know that it consists of two neutron stars separated by a distance of around 1 million kilometers. Imagine the violence of this system. Two stars, each with the mass of the sun compressed into the size of a city, spinning hundreds of times a second and careering around each other at a distance only three times greater than that from the earth to the moon. The advantage of having two pulsars for Einstein-testers is that the radio waves from one of them sometimes pass very close to the other pulsars. This means that the ultraregular radio beam passes through a region of heavily curved spacetime, which delays its transit. Careful observations can measure the delay and in that way confirm the correctness of Einstein's theory.

Another virtue of the double pulsar system is that as the stars orbit around each other, they induce ripples in spacetime that propagate outward. The ripples take energy away from the rotational motion of the pair and cause them to slowly spiral inward. The ripples have a name. They are called gravitational waves and their existence is also a prediction of Einstein's theory (they do not exist in Newtonian gravity). In one of the greatest achievements in experimental science, astronomers using the 64-meter Parkes telescope in Australia, the 76-meter Lovell telescope at Jodrell Bank in the UK, and the 100-meter Green Bank telescope in West Virginia have measured the rate at which the pulsars spiral inward to be just 7 millimeters each day, which is in accord with the prediction of general relativity. The achievement is breathtaking. These are spinning neutron stars orbiting around each other at a distance of a million kilometers and located 2,000 light-years from earth. Their behavior was predicted to millimeter precision using a theory developed in 1915 by a man who wanted to understand why two lumps of stuff dropped off a leaning tower in Pisa three centuries previously hit the ground at the same time.

Ingenious and arcane as the double pulsar measurements are, general relativity makes its presence felt here on Earth too in a much more commonplace phenomenon. The GPS satellite system is ubiquitous throughout the world, and its successful functioning depends upon the accuracy of Einstein's theories. A twenty-four-strong network of satellites circle the earth at an altitude of 20,000 kilometers, each performing two complete circuits every day. The satellites are used to "triangulate" locations on Earth using precise onboard clocks. In their high-altitude orbits the clocks experience a weaker gravitational field, which means that spacetime is warped differently for them compared to similar clocks on Earth. The effect is that the clocks speed up at a rate of 45 microseconds each day. Apart from the gravitational effect, the satellites are also whizzing around at pretty high speeds (around 14,000 kilometers per hour) and the time dilation predicted by Einstein's special theory amounts to a slowing down of the clocks by 7 microseconds each day. Taken together, the two effects amount to a net speeding up of 38 microseconds per day. That doesn't sound like much but ignoring it would lead to a complete failure of the GPS system within a few hours. Light travels around 30 centimeters in 1 nanosecond, which is 1,000-millionth of a second. Thirty-eight microseconds is therefore equivalent to over 10 kilometers in position *per day*, which wouldn't make for accurate navigation. The solution is simple enough: The satellite clocks are made to run slow by 38 microseconds per day, which allows the system to work to accuracies of meters rather than kilometers.

The faster running of the GPS satellite clocks relative to the clocks on the ground can be quite easily understood using what we've learned in this chapter. In fact, the speeding up of clocks is really a direct consequence of the principle of equivalence. To understand how it comes about, let us travel back in time to 1959 to a laboratory at Harvard University. Robert Pound and Glen Rebka have set about designing an experiment to "drop" light from the top of their laboratory to the basement, 22.5 meters below. If the light falls in strict accord with the principle of equivalence, then, as it falls, its energy should increase by exactly the same fraction that it increases for any other thing we could imagine dropping.* We need to know what happens to the light as it gains energy. In other words, what can Pound and Rebka expect to see at the bottom of their laboratory when the dropped light arrives? There is only one way for the light to increase its energy. We know that it cannot speed up, because it is already traveling at the universal speed limit, but it can increase its frequency. Remember, light can be thought of as a wave motion; a series of peaks and troughs rather like the water waves emanating outward when a stone is thrown into a still pond. The frequency of the waves is simply the number of peaks

---

* If you know that the potential energy is equal to "mgh," then you can easily see that this fractional increase is equal to gh/c^2 where g is the acceleration due to gravity and h is the height of the drop.

(or troughs) that pass a particular point every second, and these peaks and troughs can be used as the ticks of a clock. In particular, in the Pound-Rebka experiment you might imagine that Pound is sitting beside the light source at the top of the tower. He can count how many peaks of light are emitted for every beat of his heart. Now suppose that down in the basement Rebka is sitting beside an identical light source. He too can count how many peaks correspond to each beat of his heart and he should get the same answer as his colleague because they are identical light-source clocks and identical hearts. Okay, they will get exactly the same number only if they really have identical hearts, and that isn't going to be the case, but we can imagine for the sake of this argument that their hearts do beat as one. Now, let's think about how Rebka, sitting in the basement, sees the light that is arriving from Pound's light source at the top. Because the light has gained energy and thereby increased its frequency, it follows that Rebka finds that the peaks are arriving more frequently than they would if the light source were beside him. But the peaks are synchronized to his colleague's heartbeat. That means that according to Rebka down in the basement, Pound's heart would be beating faster and so he would age more quickly. The effect is a tiny one, corresponding to a speeding up of one second every 13 million years. It is testament to the skill and ingenuity of Pound and Rebka that they managed to devise an experiment capable of detecting the effect. This speeding up of time is precisely what is happening with the GPS satellite clocks. They are at a much higher altitude than the 22.5 meters of the Harvard laboratory but the basic idea is just the same: Clocks run faster in weaker gravitational fields.

Einstein's general theory of relativity, confirmed beautifully by experiment, has led us to view spacetime not as a forever-fixed blend of space and time but instead as a more dynamical entity—one that can be manipulated by the presence of matter and, since through $E = mc^2$ we know that mass and energy are interchangeable, energy too. In turn, the dynamical structure of spacetime controls the way objects move through it. No longer are we to think of space as an inert arena within which things happen and of time as the immutable and absolute ticking of a giant clock in the sky. Perhaps the most important lesson to learn in the face of this radical revision is that it is not wise to extrapolate experience beyond its realm. Why should fast-moving things behave according to the same laws as the slow-moving things we encounter in everyday life? Likewise, why should we have a right to infer the behavior of very massive objects by studying only the lighter ones?

Certainly our everyday experiences prove to be a pretty poor guide and, as Einstein has shown us, the deeper level of understanding is so much more elegant. Bringing together as it does such disparate concepts as mass and energy, space and time, and ultimately gravity, Einstein's special and general theories will stand forever as two of the greatest achievements of the human mind. In the years to come, new understanding built upon new observations and experiments may well lead to a revision in the ideas we have presented here. Indeed many physicists are already anticipating a new order in their quest for more accurate and more widely applicable theories. This humbling lesson not to extrapolate beyond the evidence is not confined to relativity—

the other great leap forward in twentieth-century physics was the discovery of the quantum theory, which underpins the behavior of all things at atomic scales and smaller. Nobody ever would have figured out how nature works at small distances based purely on everyday experience. To human beings, whose direct observations are confined to the "big things," the quantum theory is ridiculously counterintuitive, but in the twenty-first century it underpins so much of our modern lives, from medical imaging to the latest computing technologies, that we must accept it whether we feel comfortable about it or not.

Today physicists are faced with a dilemma. Einstein's general relativity, our best theory of gravity, cannot be meshed with quantum theory. Either one or both must be revised. Does spacetime "break up" at tiny distance scales? Maybe it does not really exist at all but is instead only an illusion formed by the ever-increasing set of "things that happen." Are the fundamental objects in nature tiny vibrations of energy known as strings? Or does the solution lie in some other theory yet to be uncovered? This is the frontier of fundamental physics, and those standing on the edge are both thrilled and inspired to be looking out into the unknown.

At the end of a book on Einstein's theories of relativity, it is all too easy to contribute to an unfortunate cult of personality surrounding the great man, and this is not our intention. Indeed, such a cult probably inhibits future progress because it gives the impression that science is the preserve of supermen in possession of a unique insight inaccessible to the rest of us. Nothing could be farther from the truth. Relativity was not the work of one man, although in a book about relativity this can sometimes appear to be the case. Einstein was undoubtedly one of the great practitioners of the art of science, but as we have emphasized throughout this book, he was led to his radical revision of space and time by the curiosity and skill of many. He was not a freak of nature and his intellect was not supernatural. He was simply a great scientist who did what scientists do: He took simple things seriously and followed through the consequences logically. His genius lay in taking seriously the constancy of the speed of light, as implied by Maxwell's equations, and the equivalence principle, first appreciated by Galileo.

Our hope is to have written a book that allows nonscientists to understand Einstein's beautiful theories. This understanding is within reach for nonexperts because science is really not that difficult. Given the right starting point, the road to a deeper understanding of nature is traveled in small steps, carefully taken. Science is at its heart a modest pursuit, and this modesty is the key to its success. Einstein's theories are respected because they are correct as far as we can tell, but they are no sacred tomes. They will stand, to put it bluntly, until something better comes along. Likewise the great scientific minds are not revered as prophets but as diligent contributors to our understanding of nature. There are certainly those whose names are familiar to millions, but there are none whose reputations can protect their theories from the harsh critique of experiment. Nature is no respecter of reputations. Galileo, Newton, Faraday, Maxwell, Einstein, Dirac, Feynman, Glashow, Salam, Weinberg . . . all are great, the first four were only approximately correct, and the rest may well meet the same fate during the twenty-first century.

Having said all that, we have absolutely no doubt that Einstein's special and general theories of relativity will forever be remembered as two of the greatest achievements of the human intellect, not least in the way they show how powerful imagination can be. From an inspired mix of pure thought and a little experimental data, a man was able to change our understanding of the very fabric of the universe. That Einstein's physics is both aesthetically and philosophically pleasing while also being extremely useful delivers an important lesson, the true significance of which is all too rarely appreciated. Science at its best is driven by inquiring minds afforded the freedom to dream, coupled with the technical ability and discipline to think. If the society in which Einstein flourished had decided that it needed a new power source to provide for the needs of its citizens, it is impossible to imagine that some enlightened politician would have channeled public funding into an exploration of the nature of space and time. But as we have seen, it was precisely this road that led to $E = mc^2$ and delivered the keys to unlock the power of the atomic nucleus. From the simplest of ideas—that the speed of a beam of light is one thing upon which everyone in the universe should agree—a box of riches was discovered. "From the simplest of ideas" . . . if there were ever to be an epitaph written for humanity's greatest scientific achievements, it might begin with these five words. Taking delight in observing and considering the smallest and seemingly most insignificant details of nature has led time and again to the most majestic of conclusions. We walk in the midst of wonders, and if we open our eyes and minds to them, the possibilities are boundless. Albert Einstein will be remembered for as long as there are humans in the universe both as an inspiration and an example to all those who are captivated by a natural curiosity to understand the world around them.

INDEX